In Praise of Carbon.

How We've Been Misled Into Believing that Carbon Dioxide Causes Climate Change

Fourth, Revised Edition

David Bennett Laing

This book was inspired by the principal
scientific visionary of our time,
James Lovelock

Illustrations (all from Wikimedia Commons, except as noted)

1. The carbon dioxide molecule. Image anonymous.
2. Solar and Earth radiation and the absorptivity of certain atmospheric gases. Image by Robert A. Rohde.
3. Trends in atmospheric carbon dioxide, Mauna Loa. Image by Sémhur.
4. Average monthly northern hemisphere temperature for the 20th century, NOAA. Image by David Bennett Laing.
5. NOAA record of global temperature anomalies, Image by NOAA.
6. Monthly trends in CO2, ozone depletion, and temperature anomaly for the period 1975 – 1998 Image by David Bennett Laing.
7. Infrared spectral radiation curves, Berk, A and F. Hawes, Validation of MODTRAN®6 and its line-by-line algorithm, J. Quant. Spectrosc. Radiat. Transfer, In Press, <https://doi.org/10.1016/j.jqsrt.2017.03.004>https://doi.org/10.1016/j.jqsrt.2017.03.004
8. CO_2 line spectrum, L. S. Rothman, I. E. Gordon, Y. Babikov et al., "The HITRAN 2012 Molecular Spectroscopic Database", *J. Quant. Spectrosc. Radiat. Transfer* **130**, 4-50 (2013)

9. Banded iron formation in Australia. Image by Graeme Churchard.
10. Temperature profile of Earth's atmosphere. Image Anonymous.
11. Penetration of Earth's atmosphere by three categories of ultraviolet radiation. Image by NASA.
12. Record of temperature changes over the last ice age in Greenland and Antarctica. Image by William M. Connolley.
13. Explosion of Mt. Pinatubo, Philippines. Image by Richard B. Hoblitt.
14. Non-explosive basaltic eruption in Iceland. Image by Peter Hartree
15. Effective chlorine in Earth's atmosphere Image by CSIRO Marine and Atmosphere Research (Australian government)
16. Venus, Earth, and Mars. Image by NASA.
17. Earth and Moon orbits and axial tilts, Image by NASA.
18. The carbon atom. Image by Alejandro Porto.
19. A carbon atom tetrahedrally bonded to hydrogens in methane. Image by Jynto.
20. The structure of graphite. Image by Benjah-Bmm27.
21. Various bonding styles of carbon. Image by Michael Ströck.
22. The periodic table of the elements. Image by Sandbh.
23. The chlorophyll molecule. Image by Yikrazuul.ca

24. Variation of carbon species in seawater with pH (source unknown)
25. Processes of carbon cycling in the marine environment (David Bennett Laing)
26. History of carbon dioxide through the Phanerozoic. Image by Robert A. Rohde.
27. Carbon dioxide and temperature records in the Vostok Ice Core, Antarctica. Image Anonymous.

Table of Contents

1. Carbon: A Blessing, or A Curse?

Greenhouse Warming: Fact or Fiction?

Carbon pollution, carbon tax, carbon footprint. Terms like these imply that there is something undesirable about the sixth element in the periodic table. Is such a negative attitude really justifiable? After all, everyone knows that carbon is the basis of all living things on Earth, and that it all begins when green plants take carbon dioxide from the air, and from it manufacture protoplasm, or carbohydrate, through the magic of *photosynthesis*. Those of us who aren't gifted with that magic of being able to make our own carbohydrate must consume green plants or other animals in order to get it, but without the vital input of carbon, we would soon perish.

A lesser-known fact is that every one of us consists of a little over two-thirds carbon dioxide. Most of us have been led to believe that we consist of over two thirds water, and this is true, too, but if you remove all the water, the human body actually consists of more than 67 percent carbon dioxide, dry weight. Why should we then think ill of the substance on which our very lives depend and of which we mostly consist?

In answering this question, a good place to start would be with the reservoir from which plants take their carbon dioxide, Earth's atmosphere. Two dominant gases make up the atmosphere, nitrogen, N_2, at a little more that 78 percent, and oxygen, O_2, at a little less than 21 percent, both of which are the products of biological activity. Oxygen is produced as a waste product by the great majority of photosynthetic organisms, and nitrogen is produced mainly by the denitrifying bacteria and to some extent by volcanoes. The inert gas argon, Ar, at a little less than 1 percent, is produced entirely by volcanoes.

The remaining half percent of the atmosphere, or five hundred parts per million (500 ppm), is made up of trace gases. Despite their relative scarcity, these trace gases are of great importance. Among them are certain ones that go by the name of *greenhouse gases* (GHGs), all of which have three or more atoms in their molecules. These include carbon dioxide, CO_2, (400 ppm), methane, CH_4, (1.8 ppm) and water vapor, H_2O, (highly variable, but averaging about the same as carbon dioxide, that is, 400 ppm, or 0.4 percent), Unlike the *diatomic* molecules nitrogen and oxygen, having more than two atoms allows them to absorb low intensity heat or *infrared radiation,* which is the kind that is given off by Earth's surface. This fact is well established, and it has been amply proven by many experiments. It's also well established that these greenhouse gases can re-emit heat radiation in all directions. Half of it therefore goes up and escapes to outer space, and the other half goes back down to Earth.

Where we get into trouble, however, is with the further assumption that the downward-directed component of this heat radiation that is re-emitted by carbon dioxide can cause uncontrollable warming on Earth. This hypothetical result is known by various names, the *greenhouse effect*, *greenhouse warming*, *anthropogenic global warming (AGW)*, and *climate change* among them. For the purposes of this book, I use "greenhouse warming" for this supposed effect.

The surprising fact about greenhouse warming, however, is that hard-data-based studies have never proven that it can happen. In fact, a careful search of over 10,000 climate-related, peer-reviewed journal articles by a colleague of mine revealed that only one such study has ever been done. That study was actually performed over a century ago, in 1900, and it showed, on the basis of hard data and not simply theory, that greenhouse warming doesn't happen. A more recent study that I did in 2015, and which I describe further on in this book, shows, in fact, that it can't happen.

This really shouldn't surprise us, however, because it's well known from common experience that objects, such as an apple for example, can't heat themselves with their own radiation. Planet Earth obeys exactly the same rule. This common experience has in fact been codified in the form of the second law of thermodynamics. If an object could, in fact, heat itself with its own radiation, its temperature would quickly rise to the object's melting and vaporization points, causing the object to disappear in a puff of vapor, and that quite obviously doesn't happen.

Now let's apply this important observation to the lower four to six miles (7 to 11 kilometers) or so of Earth's atmosphere, or the *troposphere*, to which all of Earth's weather is confined, and within which these greenhouse gases are well mixed. With increasing altitude in the troposphere, and thus with distance from Earth's warm surface, temperature falls at what is known as the *lapse rate* of 27.8 °F per mile (9.6 °C per km) in dry air. If the air contains water vapor, however, the lapse rate is less, because the water vapor condenses into clouds as the air rises into lower-pressure regions of the atmosphere, releasing heat of condensation as it does so. Therefore, in moist air, the lapse rate is lower, typically around 17.4 °F per mile (6.0 °C per km).

In either case, the air aloft is considerably cooler than Earth's surface below it. Now, an important principle of physics states that the *frequency*, or rate of vibration, of the radiation given off by a substance, such as carbon dioxide gas in the atmosphere, is directly proportional to the temperature of the substance, so the radiation given off by the cooler air aloft has a lower frequency than the radiation given off by Earth's warmer surface. This lower-frequency radiation is therefore quite literally powerless to raise the temperature of Earth's surface. In other words, radiation from the atmosphere is always cooler than Earth's surface, and therefore any radiation received from carbon dioxide and other gases in the atmosphere simply can't be absorbed by the warmer surfaces of the ground or sea.

Infrared astronomers are particularly concerned about any infrared back-radiation from the atmosphere that could interfere with their observations. Because of this, they choose observing locations in arid regions and high altitudes in order to avoid interference from back-radiation, but carbon dioxide isn't mentioned in this regard. It's always water vapor that they're concerned about. Significant back-radiation from CO_2 is not considered a problem. That is because they mainly confine their observations to shorter wavelengths than 13 microns, which is the upper limit of the range emitted by CO_2.

What this means is that all back-radiation from atmospheric carbon dioxide corresponds to temperatures well below any that naturally occur at Earth's surface, and consequently it is totally ineffective as a means of heating the planet.

How Radiation Interacts With Matter

Here, a brief explanation of how radiation affects matter, such as a carbon dioxide molecule, might be helpful. Carbon dioxide is a linear molecule, as is shown by the illustration below, in which a single carbon atom, having 6 protons in its nucleus, is flanked symmetrically by two oxygen atoms, each having 8 protons in its nucleus. This assemblage is surrounded by the electrons that belong to the three atoms involved. They do "double duty" by satisfying two desiderata of the molecule at once, 1. the absolute requirement that the molecule must be electrically neutral in having a combined total of 22 electrons to balance the 22 positive protons in the nuclei of the three constituent atoms involved, and 2. the desirability that the outer, or *valence*, shell of each of the three atoms has a "full complement" of eight electrons.

Since both carbon and oxygen lack enough electrons to fill their valence shells to their full capacity of eight each, the 16 valence electrons in the molecule must play the part of 24 electrons (3 x 8). They accomplish this by each of carbon's four valence electrons pairing up with one of the six valence electrons belonging to each of the oxygen atoms. This arrangement produces four strong *covalent bonds* which firmly hold the carbon dioxide molecule together. Since each such bond has two electrons, this sharing strategy enables all three atoms to have a full complement of eight electrons in its outer shell. Despite their being locked in place in covalent bonds, however, these electrons, and the molecule itself, are in constant, extremely rap-

id, oscillatory motion, the frequency of which is dependent on the temperature of the molecule. In the case of CO_2 these motions consist of symmetric and asymmetric stretching and bending.

These motions of the bonding electrons and the atoms that are bound by them involve accelerations, that is, deviations from constant speed in a straight line. It has been shown that charged particles, such as atoms and electrons, that are undergoing acceleration emit electromagnetic radiation. Thus, they give off radiation that reflects the amount of acceleration they are experiencing, which in turn reflects the temperature of the molecule. The higher the temperature, the more vigorous is the motion of the individual atoms and their bonds, and hence the greater is their acceleration.

It should come as no surprise, therefore, that any electromagnetic radiation coming in to the carbon dioxide molecule must oscillate at a higher frequency than the molecule and its electronic bonds before it can be absorbed to cause more vigorous agitation of the atoms and their bonding electrons, and a consequent higher temperature of the molecule. This, of course, is just another way of stating what we already knew, that objects can only be heated by radiation coming from other objects that are hotter than they are.

All such orbital molecular bonds, like miniature trumpets or violin strings, have certain natural frequencies of oscillation, and these are again dependent on temperature. When incoming radiation that has one of these frequencies of oscillation is absorbed, it causes a "jump" or sudden reorganization, of the orbiting electrons to accommodate the increasing separation of the atoms with increasing temperature. This is called *resonance*. Radiation that is vibrating at a frequency that is below the range of these resonant bond frequencies won't induce oscillations in molecular bonds. Radiation that is vibrating at a frequency that's higher than the lowest of the bond frequencies, on the other hand, will be absorbed.

Radiation that's not absorbed by matter is transmitted, reflected, or scattered by the matter. The only thing that this unabsorbed radiation can do, therefore, is slow the rate at which heat is lost from Earth's surface.

An analogy to this situation can be made with a blanket covering a sleeping person. Without the blanket, the person can feel cold, as his or her body heat escapes to the surrounding environment. The blanket warms up over the body beneath it, whose metabolism brings its temperature up to about 98.6 ^{o}F, but the temperature of the blanket can't rise above this normal body temperature unless it has an independent source of heat, such as is the case with an electric blanket. As with CO_2 in the atmosphere, the warmed blanket then re-radiates approximately half of its radiation downward, which retards the rate at which heat is lost from the body, thereby keeping the body warmer than it would be in the absence of the blanket, but again, no warmer than 98.6 °F.

Still, in puzzling defiance of these rather simple and obvious principles of radiative transfer, the greenhouse effect is still regarded as correct by most of the climate science community. That is, it's assumed that somehow, the downward-directed infrared radiation, even though it's cooler than the surface over which it lies, can still somehow heat up that surface. The general public also regards greenhouse theory as solid because people are regularly bombarded with propaganda from politicians and convinced scientists alike, who use this supposed "settled science" to drive their own agenda, including "saving the earth" from overheating by switching over to fuel sources that don't produce carbon dioxide. Why is this?

A Brief Excursion Back in Time

To gain some perspective on this important question, let's go back a bit in history. In the early 1860s, the great Anglo-Irish scientist and renowned Alpinist John Tyndall performed experiments in which he discovered that carbon dioxide, when exposed to infrared radiation, or heat, would absorb that radiation, and that this gas would also emit the same kind of heat radiation that it absorbed. This seemed to confirm what many people had been talking about for some time, in other words the so-called "greenhouse effect," in which certain gases in the atmosphere, like carbon dioxide, CO_2, methane, CH_4, and water vapor, H_2O, would act to warm Earth's surface like a blanket, by absorbing heat radiation given off by Earth, and by reradiating some of it back to the surface.

In 1884, the American meteorologist William Ferrell reasonably pointed out that at least, these radiation-absorbing gases should serve to slow the escape of heat from Earth to space, thereby keeping things warmer than they would be if they weren't there. In this, he was quite correct.

The three gases mentioned above are rather distinct in their distributions in the atmosphere. Both CO_2 and methane are well mixed, although methane is comparatively rare. CO_2 is about 4600 times as concentrated as methane, and the lifetime of methane in the atmosphere is about 1/16 as long as that of CO_2. While these two gases are rather constant with regard to their concentrations and residence times, water vapor is highly variable in terms of its concentration, or its *absolute humidity*, and its residence time. The maximum absolute humidity of water vapor in natural saturated air at about 86 ^{o}F (30 ^{o}C) is about 4.3 percent, or about 100 times that of carbon dioxide, but at -58 ^{o}F (-50 ^{o}C), it is close to zero percent.

As the figure below shows, these gases behave differently with respect to how they absorb Earth's infrared heat radiation. Notice that in this commonly used diagram the scale of heat radiation, shown in blue in the figure, has been greatly exaggerated relative to Sun's output. Further, the rendering of atmospheric absorption as percent (gray bars) gives a false impression of the importance of absorption in the infrared. If absorption were expressed as actual amounts, the height of the gray bands would diminish in going from left to right in the diagram. In reality, Earth's radiation only amounts to a tiny fraction of Sun's spectrum, shown here in red.

Because of its comprehensive absorption of infrared radiation over a broad range of wavelengths, it's quite clear that water vapor is a far more effective greenhouse gas than carbon dioxide. A humid atmosphere can clearly absorb most of Sun's infrared heat as well as most of Earth's, and the cloudiness that usually accompanies a humid atmosphere can also reflect away most incoming solar radiation, including that which has higher frequencies than infrared, as is well-known from common experience. On the other hand, in a dry atmosphere, there is no such reflection, and practically all the absorbing is done by carbon dioxide and methane.

Because of water vapor's great absorptivity, it's far more important than carbon dioxide in determining the radiative balance of the atmosphere, despite the evanescent nature of its comings and goings with the movement of weather systems across the globe. Nonetheless, the evident relationship between carbon dioxide and warming captured the interests of nineteenth

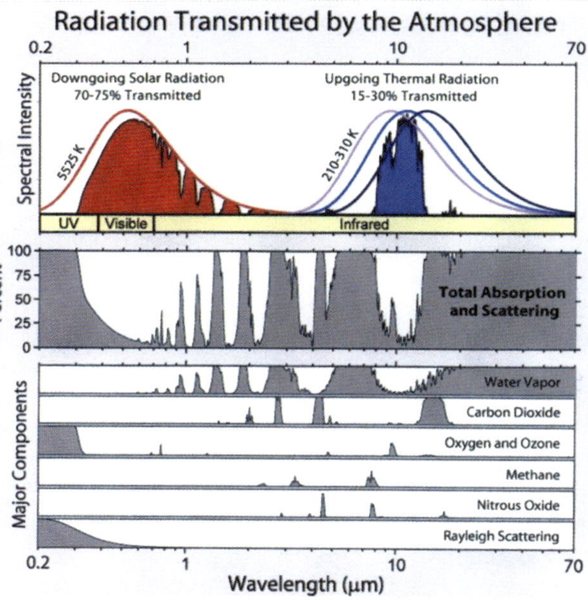

century scientists, and the quantification of this relationship became somewhat of a holy grail in the latter decades of that century.

The prominent Swedish physicist and chemist Svante Arrhenius rose to that occasion, and in 1896, he calculated how much infrared radiation from Earth should be absorbed by carbon dioxide in the atmosphere, and from this, he established his well-known equation for how the surface temperature on Earth could be increased arithmetically by a geometric increase of carbon dioxide in the atmosphere. The rather simple Arrhenius equation, still in use in today's sophisticated computer modelling, essentially says that a doubling of carbon dioxide in the atmosphere will result in so many degrees of increase in atmospheric temperature.

In Arrhenius's time, the contemporary concentration of CO_2 in the atmosphere was about 280 parts of per million (or *ppm*), a value that had held fairly steady since before the industrial revolution, and, in fact, ever since the end of the last ice age. During the ice age itself, which lasted about a hundred thousand years, the concentration of CO_2 in the atmosphere had fallen to as low as 180 parts per million simply because the oceans were colder then, and colder water can hold more CO_2 in solution than warmer oceans can, so they absorbed up to 100 parts per million of CO_2 from the atmosphere.

According to Arrhenius's formula, the rise in atmospheric temperature that should result from a doubling of atmospheric carbon dioxide should be somewhere between 3 and 8 ^{o}F (1.67 and 4.4 ^{o}C). To give some idea of the magnitude of such an effect, the temperature rise from the last ice age to modern conditions was about 12 °C, or more than two and a half to seven times this amount.

Then, in 1900, Arrhenius's contemporary and countryman, atmospheric physicist Knut Ångström (pronounced *Ong*-strerm) thought it would be a good idea to test Arrhenius's compelling theory with an actual experiment, a necessary step for the validation of any new scientific theory, and here, I'll digress for a moment to make an important point about theories and their testing.

A theory is basically a well-accepted statement of truth about the natural world. The theory of evolution is one such, and its general acceptance is based essentially on the fact that no test of that theory, that is, no observation from Nature or laboratory experiment, has ever shown it to be incorrect, despite there having been countless numbers of such observations and experiments.

Nonetheless, just one observation or experimental result that is inconsistent with the theory could call it into question. If that new evidence, that new observation or experiment, were then repeated with the same negative results, the theory of evolution would have to go back to the drawing board for revisions that would bring it into line with the new evidence. If, however, it couldn't be altered to bring it into conformity with the new information, it would have to be rejected as inadequate to explain everything that it sought to explain in Nature, and a new theory would need to be developed in its place.

Greenhouse warming is another such very well accepted theory. It developed around Tyndall's investigations and Arrhenius's theoretical formula, and it was generally hailed by scientists and by the scientifically literate public of the latter half of the nineteenth century as sort of an official confirmation of what they had thought all along.

The fact is, however, that any theory is nothing more than a tool in the investigation of truth. It should not be regarded as an automatic codification of reality unless it has been supported by a significant body of observations and experiments with *all* of which it is fully consistent.

Ångström, the leading atmospheric physicist of the turn of the twentieth century, realized the importance of this principle, and he duly accepted the responsibility of testing Arrhenius's new theory with a series of very carefully designed experiments at various altitudes, from which he drew the unexpected and surprising conclusion that there was in fact very little effect on atmospheric temperature from a rise in the concentration of atmospheric carbon dioxide.

This inconvenient result infuriated Svante Arrhenius, who tried vigorously to defend his theory and his famous equation, but to no avail. Ångström's results, based as they were on hard data from the Earth system, were considered unarguable, and they were rightly respected, as was he, by the scientific community.

What is clear here is that Nature is always right. Hard data from the real world are the ultimate repositories of truth. Human beings, on the other hand, are often wrong, and in order to be considered as trustworthy representations of the real world, their brilliant ideas about how the world really works—in other words, their hypotheses, theories, and laws—must always be assessed in relation to hard data from Nature.

Ångström's careful series of experiments in 1900 did just that—they put Arrhenius's greenhouse warming theory to the test by clearly demonstrating that an increase in the carbon dioxide content of air in a specially constructed glass cylinder didn't result in increased temperature when exposed to a source of infrared radiation of a frequency that is absorbed by the carbon dioxide. This relation held equally at sea level and at the summit of a 10,000 foot mountain.

Despite Arrhenius's stature, and his loud protestations, Ångström's careful experiments were considered conclusive by the climate community of the day, a fact that reflects his equal stature to Arrhenius as an atmospheric scientist. As a result, the theory of greenhouse warming was put to rest for thirty-eight years.

Then, in 1938, a British steam engineer and climate hobbyist (!) named Guy Callendar wrote the first of a series of very persuasive articles arguing that greenhouse warming should work as Tyndall and Arrhenius supposed. In fact, he wrote thirty-five such articles on this and related topics until his death in 1964, and these succeeded in reigniting scientific interest in the concept. In the United States, the Canadian researcher and Texas A&M University professor Gilbert Plass became a part of this revival, providing the first computerization of the concept, and he predicted that a doubling of atmospheric carbon dioxide would produce a warming of 3.6 °C.

Since the late 1960s, Syukuro "Suki" Manabe of Princeton University and NOAA's Geophysical Fluid Dynamics Laboratory has worked with Dr. Kirk Bryan, Jr. and others to develop an increasingly sophisticated series of computer models with increasingly finer resolution. These models were designed to mimic the general circulation of Earth's atmosphere under various influences. Among these influences was carbon dioxide. The validity of Arrhenius's equation and theory about this greenhouse gas was simply assumed by these workers, largely due to Callendar's efforts.

I went to a conference on atmospheric modeling in Pasadena, California in 1990, where I attended a packed lecture by Manabe, who was talking about his latest modeling efforts. These purported to show a cooling of the stratosphere because of increased absorption by carbon dioxide in the troposphere. Along with this was a concomitant warming of the troposphere, particularly in the northern portions of the northern hemisphere. At the end of the lecture, during the question and answer session, I asked Dr. Manabe why it was that his computer model was predicting maximum tropospheric warming in the high latitudes of the northern hemisphere when actual records showed that maximum warming was occurring in the low latitudes of the southern hemisphere. For a moment, he stroked his chin, then he raised one finger and said, "Bad data. Next question, please."

This was actually not a bad answer for him to have given, since in fact, the data were indeed bad because a significant portion of them had been derived from water scooped up by wooden buckets on shipboard. Nonetheless, blaming bad data for discrepancies between models and reality is, at best, a rather shaky excuse, and it is highly unlikely that bad data alone could account for such a glaring discrepancy from reality in the models.

Meanwhile, the Air Force Cambridge Research Laboratories were compiling the very extensive HITRAN (HIgh resolution TRANsmission) database of radiative absorption and emission by atmospheric gases, including carbon dioxide. Many ardent adherents to greenhouse warming theory, and in fact even many climate scientists, regard the HITRAN database as de-facto proof of warming, whereas it is actually anything but that because it deals only with absorption and re-radiation by greenhouse gases, in exhaustive detail, to be sure, but in truth it says nothing at all about warming resulting from absorption of re-radiation by Earth's surface.

In short, in the latter part of the 20th century and the first part of the 21st, much hard work has clearly been done on the subject of greenhouse gases, but all this did absolutely nothing to validate greenhouse warming *theory*. It seems abundantly clear that the theory has not been tested experimentally or by observation since Ångström's non-supportive experiments in 1900. As I mentioned early in this chapter, this was revealed recently by a colleague of mine, Dr. Peter Ward, a volcanologist now retired from the US Geological Survey in a fruitless search of over 10,000 climate-related journal articles, for any hard-data study of CO_2/warming.

Astonished by this finding, I then added a study of my own in 2015, using contemporary climate data, which had the interesting result of coming to the same general conclusion that Ångström's 1900 study had, namely, that there was very little effect. In other words, the tenets of the theory of greenhouse warming today rest entirely on concepts that are not backed up by hard data.

For scientific investigation, this is simply unthinkable. Since the incident in Pasadena with Manabe, I had assumed a cautious stance with respect to greenhouse warming, but I had no idea that the entire theory was tested by only one experiment, and that the conclusion of that experiment was that the theory was incorrect! Since those discoveries, I have read a few articles that have gone to considerable lengths to claim that Ångström's experiments were in various ways flawed, including some inferences that his laboratory assistant Koch was in some way incompetent, but that the best atmospheric physicist of the day would have hired an incompetent assistant for an important, published test of theory, and that he would have neglected to check Herr Koch's work before publishing his results is quite simply beyond credibility.

An Attempt to Get at the Truth

For my own analysis, I used available real data from the Earth system, restricting my investigation to the northern hemisphere for a number of reasons. First of all, the bulk of available, fairly reliable, and relatively long-term climate data are from this hemisphere. Second, the tropospheric circulation covers the surface area in a more representative fashion than in the southern hemisphere, where large expanses of ocean water favor a strictly zonal, or east-west, flow, thereby limiting coverage to certain bands paralleling the equator. Finally, most of the effects of ozone depletion in the southern hemisphere are confined to the well known "ozone hole" over Antarctica, which is essentially well sealed off and hence relatively isolated from the more northerly regions of the southern hemisphere by the south polar vortex.

I started with the so-called "Keeling curve" of carbon dioxide from Mauna Loa on Hawai'i. which records the atmospheric increase of that gas since the late 1950s. As the illustration below shows, this curve displays a series of regular "squiggles" (red), representing the annual cycle of biological decay and photosynthesis, superimposed on the general upward trend of increasing CO_2. This annual cycle amounts to an average annual variability in CO_2 of some six ppm superimposed on the increasing long-term trend of the Keeling curve over the available record. The illustration here also shows an inset, which represents this annual cycle of CO_2 variation. This inset removes the long, upward trend of the blue line and shows just the mean month to month changes through the yearly cycle.

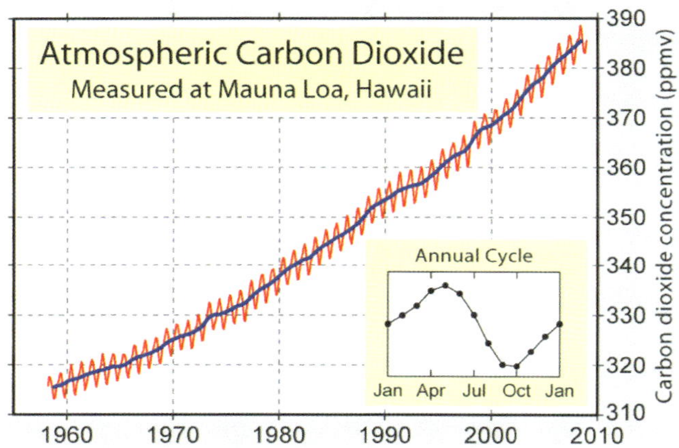

The rise to a peak in May in the inset represents the average release of CO_2 by vegetation that died in winter and decayed in spring, whereas the descent to a trough in September and October represents the removal of CO_2 from the atmosphere by photosynthesis over the summer, respectively. Of course, this applies only to the northern hemisphere. Records from similar curves of CO_2 in the southern hemisphere have their peaks and troughs reversed from those that appear in this graph.

I reasoned that if in fact variability in CO_2 did have an actual effect on temperature, then this average annual cycle of CO_2 should show up in the record of temperature anomalies from the northern hemisphere. The temperature anomaly record, however, would also have to be on a month-to-month basis in order to show any monthly changes caused by the variations in the Keeling curve.

Furthermore, in order to focus only on these monthly changes, any long-term trends, such as *insolation*, or solar input, would have to be removed from these monthly values in temperature. That condition could be satisfied if the monthly temperature anomalies were presented as deviations from averaged monthly values over an earlier base period.

Fortunately, just such a record was readily available from NOAA. The curve below shows the averaged monthly 20th century base curve from which these monthly anomalies were calculated. It is a very smooth, sinusoidal curve, showing no monthly effects at all except for the very obvious one of insolation, which, in the northern hemisphere, peaks in summer (July) and is at a minimum in winter (January). This reassured me that no long-term effects were present in the monthly temperature anomalies that I used in my analysis.

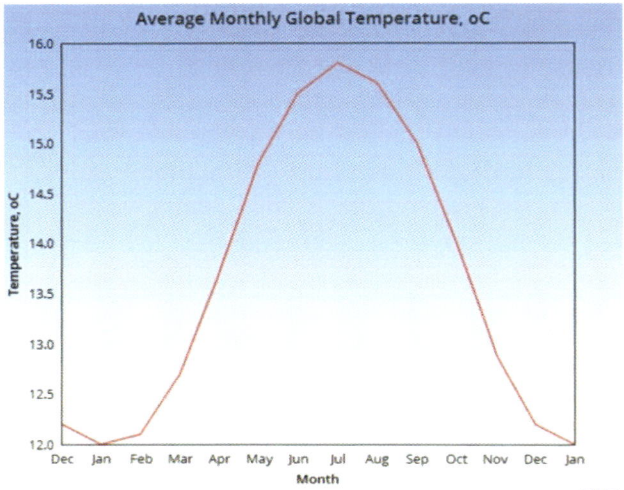

NOAA

For the study, I chose the 24-year period 1975 to 1998 because during this period global temperature shot up by about 0.8°C before levelling off in what has been called the "global warming hiatus" or "pause" (see the right side of the NOAA curve below showing the global temperature anomaly from 1880 to 2014). I took this sudden rise between 1975 and 1998 to indicate that something important happened then to affect global temperature, although I reserved judgment on the generally accepted conclusion that it was due to an increase in atmospheric carbon dioxide.

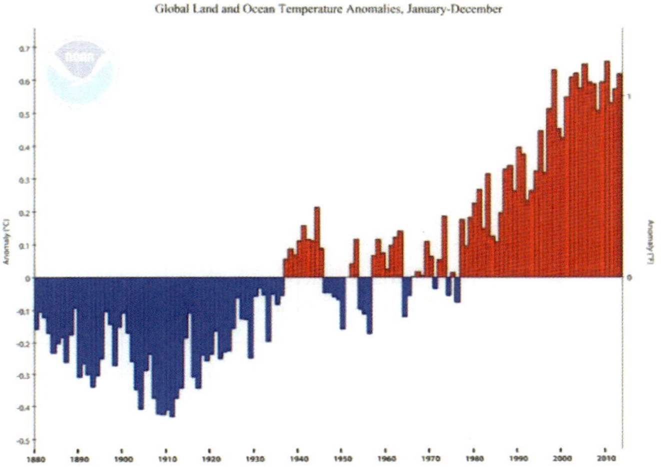

The names "hiatus" and "pause," incidentally, somewhat presumptuously suggest that something unexpected has interfered with the accepted computerized pattern of warming by carbon dioxide increase, and that Earth has therefore done something wrong!

In the graph below, which I constructed using the northern hemisphere data, the red curve for the NOAA temperature anomalies represents the monthly averages for the 1975 to 1998 analysis period. The blue curve in the graph is carbon dioxide, and it is taken from the inset in the Keeling curve graph. I converted the actual monthly values of the two curves into percentages (the highest value is 100%, and the lowest is 0%) so that I could show both curves at the same scale on the same graph. This is a standard statistical technique which doesn't affect the data in this conversion, since it results simply in a comparison between the two curves.

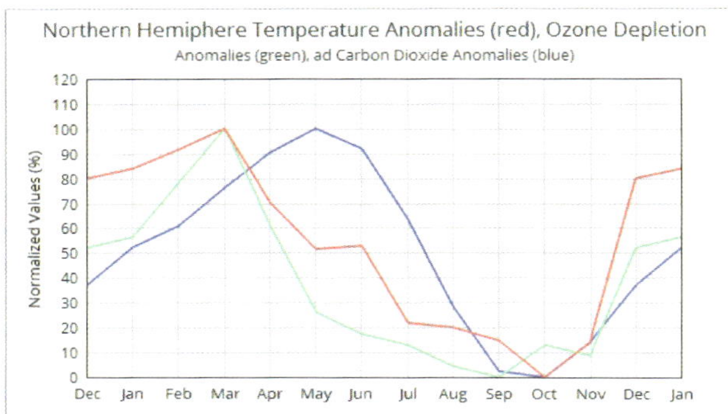

If indeed CO_2 had some effect on temperature, then I would have expected the red curve to peak in the same month as the blue CO_2 curve or perhaps in the following month of June. The temperature curve, however, actually peaks in March, two months *before* the CO_2 curve! The only possible interpretation of this is that CO_2 couldn't possibly influence temperature because its maximum, or peak, occurs two months *after* the temperature peak. One could perhaps argue that for some reason, there's a ten-month lag between peaks in CO_2 and temperature, but such long delays are uncharacteristic of the very fluid and quickly overturning troposphere, in which things usually take place fairly fast.

A slight upward deflection does occur in the red curve in the month of June, however, following the May peak in CO_2, indicating that there may be, in fact, some minor influence from CO_2 on temperature, but it's so small as to be insignificant. The correlation coefficient between these two curves is rather low, 0.54, indicating a low likelihood of a functional relationship between them.

Why, however, should the temperature anomaly curve reach a maximum in March, of all months? Here is where the third, green, curve in the graph comes in. It represents the record of the stratospheric ozone anomalies from Arosa, Switzerland. I added ozone depletion to this analysis because my colleague Peter Ward, while he was working with the Greenland ice cores, had proposed on the basis of his findings that chlorine emitted in the form of hydrogen chloride (HCl) by the non-explosive, basaltic volcanoes on nearby Iceland might prove to be a potent source of global warming. I inverted this curve so that instead of showing anomalies of ozone concentration, it actually shows the *depletion* of the ozone layer in the lower stratosphere.

Ward's concept is explained in greater detail in Chapter 2, but a brief resume of his idea is as follows. After its release, volcanic HCl is swept upward by winds to be deposited in *polar stratospheric clouds* (PSCs) that form in the stratosphere in mid-winter. There, the HCl becomes *photodissociated*, or broken down, in early spring (March) by sunlight returning to the northern hemisphere. This process releases monatomic chlorine atoms, which react with and destroy ozone catalytically. This, in turn, thins Earth's ozone layer, which allows a greater than normal influx of high-intensity solar ultraviolet-B (*UV-B*) radiation to reach Earth's surface. Under normal conditions, most of this radiation is consumed in the photodissociation of ozone in the stratosphere. More UV-B radiation reaching Earth's surface through a thinned ozone layer should cause significant global warming. I felt that I should test this possibility along with CO_2 in my analysis.

As with the temperature anomalies, I also presented the inverted ozone figures in the form of anomalies from averaged monthly values over a base period in order to remove any long-term trends. Clearly, not only does this inverted ozone anomaly curve peak in the same month (March) as the temperature anomaly curve, but both curves peak sharply together in that month.

Both these facts suggest that the two curves are related. Nonetheless, they don't prove that they are, because *correlation*, no matter how good it is, doesn't ever imply *causation*. If it did, one could argue that since the Dow Jones Industrial Average went up over the same time interval during which global temperature anomalies have been increasing, that means that the DJIA caused the increase in temperature! Depressingly enough, claims of this sort have actually been made. The correlation coefficient between these two curves of ozone depletion and temperature anomalies is high, 0.92, which also suggests, but doesn't prove, a causal relationship.

In summary, the conclusions that I drew from this graph were 1) that because the peak in annual variation of CO_2 in May occurs two months after the March peak in annual variation of the NOAA temperature anomaly curve, CO_2 couldn't have anything but an insignificant effect on temperature anomalies, shown by a slight uptick in the red curve in June, and 2) that ozone depletion *could* have a significant effect on temperature because its annual peak coincides sharply with the annual peak in the NOAA temperature anomaly curve in the same month (March) in which ozone depletion is known to be most active in the northern hemisphere. This would happen because in that month sunlight returns to the northern hemisphere to liberate ozone-destroying chlorine that has been entrained within high, polar stratospheric clouds. Importantly, the chlorine-temperature anomaly relation would explain why the temperature anomaly peaks in March, an otherwise inexplicable circumstance.

In the best of all possible worlds, my analysis would have been welcomed with open arms by the scientific community, and I would have had no trouble getting it published in a leading peer-reviewed journal, but this, unfortunately, is not the best of all possible worlds. I went through the motions of submitting the research to three such journals, which summarily rejected it without comment, as I rather expected they would. Of course, each new submission requires a re-write according to the specific requirements of the journal to which the article is submitted, so I soon tired of this game and decided instead to self-publish my work, which has thus far proven fruitful, but then an unexpected invitation from the Multidisciplinary section of the Asian Academic Research Journal led to publication in the September, 2018 edition of that journal The url is: www.asianacademicresearch.org/2018_paper/september_md_2018/3.pdf.

True Believers and the Politics of Climate Change

Beyond the apparent lack of data-based and observation-based evidence supporting the theory of greenhouse warming within the climate science community, and the general unwillingness to consider other scientifically reasonable explanations for the observed phenomenon of warming, I began to notice, with my increasing involvement with the issue, that the people who were promoting the concept of greenhouse warming were in general quite aggressive, defensive, and cocksure, which I had seen in other cases before. I had always taken this as a sign of insecurity in people who were not quite sure of what they were proposing, and I saw no reason to assume otherwise in this case.

Particularly in view of the lack of data, it seemed clear to me that the people with whom I communicated on the subject were circling the wagons around a concept that lacked adequate proof. Most of the arguments that I heard were more political in nature than scientific. For example, proponents would claim that there was a "consensus" among climate scientists. Consensus, of course, is central to politics but it has no relevance whatever to science, where either something is right or it's not, and no degree of consensus can change a concept from wrong to right. Even worse, people were saying that "the science is settled." Of course, nothing in science is ever "settled." Science is all about maintaining an open-minded approach, and all the minds I encountered seemed to be quite closed with regard to this issue.

On top of this was a clearly political divide between so-called "warmists" and so-called "deniers," which I thought was rather silly, but it did heighten my sense that the debate was anything but scientific. Personally, I've always thought of myself as a progressive or a liberal, but I do have some conservative attitudes as well. All of a sudden, however, because of my questioning of the validity of greenhouse warming, I found myself firmly in the same camp with the Koch Brothers and Big Oil, against both of which I've signed many petitions seeking to redress various injuries that I felt their policies would cause for the natural environment and the American people. Beyond this, I have no financial investments whatsoever, and consequently I have no financial interests in the petroleum industries. My interest in this issue is the purely academic one of trying to use sound scientific principles to arrive at a correct view of this aspect of reality.

What is even more troubling is that I find that those who militantly support the idea of carbon dioxide's role in global warming are generally also very much in favor of globalism and free trade. I most emphatically oppose these because they favor big, multinational corporations over the interests of the citizens of our nation.

Furthermore, the findings of my analysis run counter to the pronouncements of the Intergovernmental Panel on Climate Change (*IPCC*) located at the World Meteorological Organization (WMO) headquarters in Geneva, Switzerland. This body, created in 1988 by the WMO and the United Nations Environment Programme (UNEP), and sanctioned by the United Nations General Assembly, is essentially a clearinghouse for scientific information on climate change. Its mission is to select and disseminate scientific information on this subject to interested parties across the globe, and to assess the geopolitical implications of such information. The IPCC itself is therefore not a scientific organization, although it is largely staffed by scientists. It only selects, processes, and evaluates scientific information. Notably, it "produces reports that support the United Nations Framework Convention on Climate Change (*UNFCCC*), an international treaty on climate change, whose ultimate objective is to "stabilize greenhouse gas concentrations in the atmosphere at a level that would prevent dangerous anthropogenic interference with the climate system," a policy that clearly accepts as a given the conclusion that warming results from an increase in anthropogenic atmospheric carbon dioxide concentration.

In other words, the IPCC supports a political agendum, and must therefore be regarded as primarily a political body. Is this really "science in the public interest?" I have always seen such collusion between scientists and politicians as suspect, because it favors those scientific conclusions about reality that support certain political ends above others, whereas science really should be strictly impartial and apolitical. In this case, that agendum appears to be globalism, the imposition of world government upon nation states, and what could be a more compelling case for doing this than the urgent need to replace the very energy source that created our modern civilization with sources that don't inject carbon dioxide into the atmosphere? I'll have more to say about all this in the final chapter.

In any case, my attempts to present a fresh idea, which should be welcomed in science, have met with more than the usual resistance from professionals who have devoted significant amounts of time and research to more orthodox views, and who have been paid significant amounts by funding sources that see value in supporting orthodox views.

Why Carbon Dioxide Can't Possibly Be a Greenhouse Gas

All of the hard-data observations given above tell us quite clearly that carbon dioxide can't be a greenhouse gas, but they don't really tell us why this is the case. The answer to this is really quite simple. As noted above, the heat given off by any material object results from the molecular vibrations of its surface. In the case of Earth, virtually all of this comes from Sun, largely in the form of ultraviolet and visible light. Because of Earth's great distance from Sun, its surface receives only a small portion of this radiation, all of which it must give back on a daily basis in the form of infrared emissions to space. The heat is generated, as I mentioned, by the molecular bonds of Earth's surface, which vibrate resonantly with incoming solar radiation, and in so doing, generate infrared radiation of their own. This infrared output is released to space at the same rate, in terms of energy, as incoming solar radiation, thus preventing a buildup of heat energy on the planet.

Enter here Joseph Fourier, who, in 1824, suggested that carbon dioxide and other polyatomic trace gases in Earth's atmosphere might act as "greenhouse gases," absorbing and reradiating outgoing energy from Earth, and raising Earth's temperature with the half of the radiation that was directed back to Earth. This idea became very popular, and when John Tyndall tested it in 1864, proving that CO_2 did, indeed, absorb radiation, this seemed to be a vindication of the idea, which made people quite happy, despite the fact that no one had yet shown that back-radiation from CO_2 could actually be absorbed by Earth's surface.

When Arrhenius came up with his formula in 1896 quantifying the "greenhouse effect," that seemed to put a stamp of officialdom on the concept, and when Ångström disproved it in 1900, the people were clearly disgruntled, but Guy Callendar saved the day with his persuasive articles, and today, CO_2/warming again reigns supreme.

The trouble is, the peer-reviewed literature still lacks proof of the concept. Why? The necessary experiment is quite straightforward, All that's needed is a design showing that receiving matter (Earth's surface) actually absorbs back-radiation from atmospheric CO_2. One small detail, however, is that care must be taken that the temperatures of the transmitting source and of the receiving sink must be natural, that is, equal to what they would be in the real world. Why is this step so necessary?

For the answer to this, we must retreat to the microworld of the molecules of Earth's surface. There, we find that, as noted, the vibrations of the frictionless molecular bonds that make up the surface is a function of the frequency and the amplitude of the radiation being emitted by Sun. These vibrations determine the temperature of the surface, and thus the frequency of the emitted radiation. Thus, these frictionless bond vibrations must be of frequencies that will produce the temperature curves in the following graph. Note that the amplitudes of the vibrations are greatly reduced by Earth's great distance from Sun.

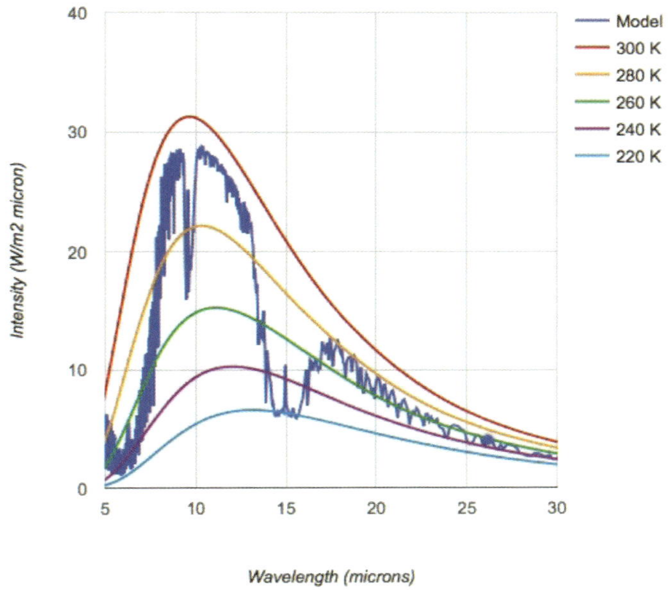

The colored lines in this graph represent the range of temperatures that are typically experienced at Earth's surface, and the ragged, blue curve superimposed on them represents an actual trace of radiation recorded by a satellite 70 km above the tropics. Notice that there are several "bites" taken out of this blue curve, the major one on the left from 5 to 7 microns is from water vapor, as are the series of smaller "bites" between 17 and 30 microns on the graph's right side. The icicle-like bite centered on about 9.7

microns is due to ozone, and the big bite between 13 and 17 microns is due to CO_2.

Notice that all these bites are taken entirely out of Earth's outgoing infrared radiation. Since back-radiation from greenhouse gases must be essentially the same as ingoing radiation, this means that the frequencies of ***all*** the back-radiation in GHGs are ***already present*** in the vibrations of the molecular bonds, and therefore that back-radiation can't induce the bonds to vibrate any faster than they're vibrating already. This in turn means that back-radiation from GHGs is ***totally ineffective*** in increasing the vibrations of the molecular bonds, and thus the temperature, of Earth's surface. Since the bonds can't use this back-radiation, the surface doesn't absorb it, but simply scatters, reflects, or transmits it. This is the all-important key to why greenhouse gases are quite ineffective in changing the temperature of Earth's surface.

There are other reasons as well. The following graph shows that portion of the CO_2 spectrum that is actually strong enough to affect Earth's atmosphere. First, this is a ***line spectrum*** in which the constituent frequencies are separated into discrete, vertical lines spaced apart by fixed, integral amounts. Each line represents a frequency at which one of the two bonds in CO_2 is vibrating, or at an overtone thereof. These so-called Ritz-Rydberg lines are usually interpreted as quantum differences in energy, but they can also be interpreted to represent resonant overtones, which also have integral spacing, and I find this a lot more convincing.

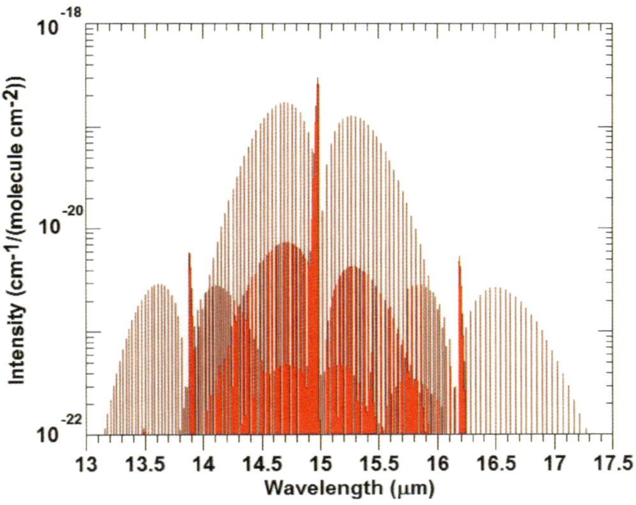

The range over which CO_2 absorbs and emits is represented by either side of the graph, and its maximum is clearly shown at 14.95 microns. At lower levels in the atmosphere, the spectral lines are compressed so that this distribution of CO_2 radiation approximates a continuous Planck distribution, in which case the maximum vibration at 14.95 microns serves as an approximation to a Planck spectrum's "Wien temperature," i.e., the actual temperature of the transmitting object, in this case approximately -79 °C. Theoretically, it's a bit more shaky to compare the bounding frequencies of 13 and 17 microns to actual temperatures, but if we do that (and there's some justification for doing so beyond the fact that it's often done), the resulting temperature range is from -51 to -103 °C. This temperature range, and its central "Wien" value correspond to a range of temperatures well below those usually experienced at Earth's surface except for occasional cold snaps in winter at the South Pole.

What this means is that back-radiation from CO_2 is much colder than the range of temperature normally encountered on Earth's surface, and as is well known, cooler objects (here, CO_2), can't transfer heat to warmer ones (here, Earth's surface). So from both of the perspectives just discussed, CO_2 back-radiation is unable to add heat to Earth's surface, which disqualifies it as a greenhouse gas.

2. About Ozone

The Ozone Shield: An Earth Exclusive

Since the discussion in Chapter One identified ozone depletion as a likely cause of global warming, a brief excursion into the role of ozone might be appropriate before taking a more in-depth look at carbon as an essential component of the Earth system.

Earth is unique among the bodies of the Solar System in having a significant component of oxygen in its atmosphere, accounting for 20.95 percent of its gaseous components. Nitrogen makes up most of the remainder, at 78.09 percent. Both these gases have their origins in ongoing biological processes, photosynthesis in the case of oxygen, and bacterial denitrification in the case of nitrogen. Because of its unique oxygen component, Earth also has a second unique feature, a diffuse layer of *ozone* in its lower stratosphere. Normal oxygen gas consists of molecules that are pairs of oxygen atoms, O_2, whereas ozone consists of three atoms of oxygen, O_3. This diffuse layer of ozone is referred to as an "ozone shield" for good reason because it shields Earth from some of the highest energy radiation in the solar spectrum.

Probably within the first half billion years of Earth's 4.54 billion year history, living organisms, mainly cyanobacteria (formerly incorrectly known as "blue-green algae") evolved the photosynthetic energy pathway. With this very efficient system, the cyanobacteria were able to split water molecules, stripping hydrogen ions from them and adding them to carbon dioxide to create protoplasm, composed largely of *carbohydrate*, and releasing oxygen into the atmosphere as a waste product. The chemical reaction looks like this:

$$H_2O + CO_2 + h\nu \rightarrow (CH_2O) + O_2\uparrow.$$

(Water plus carbon dioxide plus solar radiation yield carbohydrate plus oxygen gas.)

Notice that hν, a shorthand expression for energy from sunlight, is necessary to make the reaction go from left to right, something it doesn't do in the absence of some kind of energy source, such as sunlight. In hν, h is Planck's constant and ν (the Greek letter "nu") represents the frequency of radiation.

Up until about 2.5 billion years ago, most of the waste oxygen produced by this reaction combined chemically with reduced materials at Earth's surface, such as water-soluble *ferrous iron*, Fe^{2+}, dissolved in the oceans. This was oxidized into insoluble ferric iron oxide *hematite*, Fe_2O_3, in which ferrous iron has been totally converted to *ferric iron*, Fe^{3+}. Ferrous iron could also be partially oxidized into the magnetic oxide *magnetite*, Fe_3O_4, in which two thirds of the Fe^{2+} is so converted. This oxidized iron was laid down mainly in the form of *banded iron formation*, usually consisting of dark layers of iron oxides interbedded with reddish shale or chert. The illustration below shows a spectacular outcrop of banded iron formation in Australia.

By about 2.5 billion years ago, this process had used up most such reduced phases at Earth's surface. This accounts for the general absence of reduced materials in the geological record from that time on. After this reduced material was exhausted, the photosynthetic activity of the abundant cyanobacterial colonies living in the protected shoreline environments of the time began to deliver oxygen voluminously to the atmosphere instead.

Earth presumably hadn't had a stratosphere before this, but the presence of a high percentage of free oxygen in the atmosphere provided the necessary conditions under which a stratosphere with an ozone shield could form. This is the case today, and we can therefore infer that the same thing would have happened 2.5 billion years ago. This is how:

1) Near the top of the stratosphere, at altitudes approaching 50 km., intense ultraviolet solar radiation that has a frequency of at least 1250 *terahertz* (THz), or 1250 quadrillion cycles per second, photodissociates the normal diatomic oxygen and nitrogen molecules of the atmosphere into single atoms. This process generates heat, which accounts for the fact that the *stratopause* or top of the stratosphere, at about 50 kilometers altitude, has a temperature of about 0 ^{o}C, which is only about 15 ^{o}C below the temperature of Earth's surface. This is shown by the curve of the US Standard Atmosphere in the figure below. The monatomic oxygen atoms thus produced are highly reactive, and therefore they combine with normal diatomic oxygen molecules, O_2, in the atmosphere to form ozone, O_3.

2) Being 50 percent heavier than oxygen, ozone settles to a level near the bottom of the stratosphere, forming the "ozone layer." Here, slightly less intense ultraviolet solar radiation that has a frequency of at least 96 THz photo-

dissociates the ozone molecules into normal diatomic oxygen and monatomic oxygen, again releasing heat in doing so.

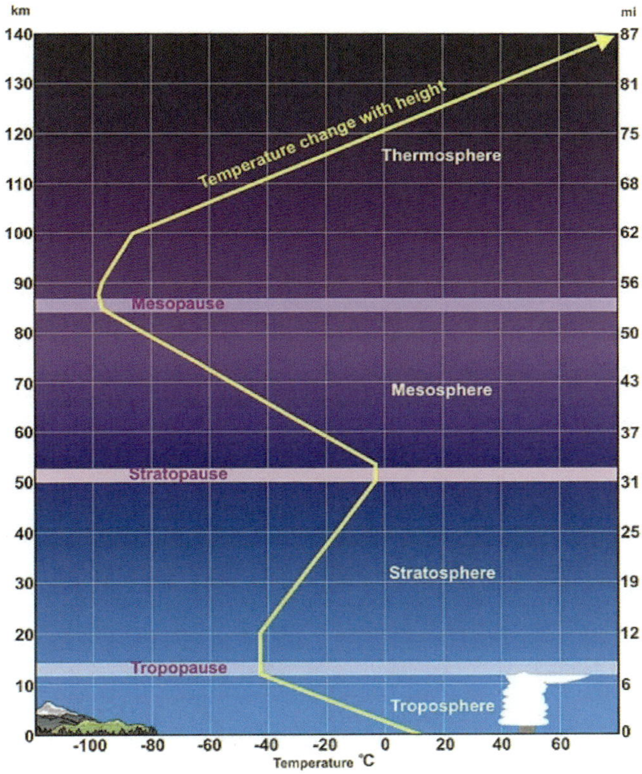

Because ozone is most concentrated near the base of the stratosphere the process of ozone destruction is most active here. The diagram below shows how the three categories of solar ultraviolet radiation, UV-C, -B, and –A, decreasing in frequency in the order given, are progressively removed from Sun's radiation spectrum as they interact with diatomic oxygen, nitrogen, and ozone in the stratosphere. Only UV-A and a small amount of UV-B normally survives these processes to reach Earth's surface, but when the ozone layer is thinned, then a variable amount of UV-B can get through to that surface, where it is absorbed by water, resulting in global warming.

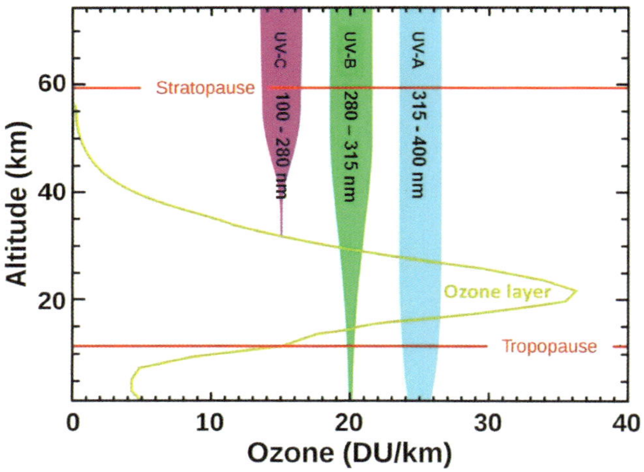

In Earth's underlying troposphere, in contrast, temperature drops with altitude above Earth's Sun-warmed surface because such heat-generating photodissociations are not present there, except when anthropogenic ground-level ozone pollution is significant and UV-B penetration is sufficient. Since surface air is warmed by Earth, it has a tendency to rise, causing turbulence, hence the name "*troposphere*" for this lowest part of the atmosphere, meaning the "sphere of overturn." In the stratosphere, however, temperature tends to increase with altitude, due to the photodissociation of ozone and diatomic oxygen. This confers great stability on the stratosphere, since air has no tendency to rise under these conditions, in which the lighter, less dense air is on top. Therefore, all convective activity in Earth's atmosphere—the source of Earth's weather—happens below this in the troposphere, which extends to 6 to 10 km above Earth's surface.

The photodissociations, however, progressively use up Sun's most intense, high frequency radiation before it can reach Earth's surface, as the above figure shows. Ultraviolet-C (magenta), ranging between frequencies of about 3000 and 1070 terahertz, is completely absorbed, or used up in photodissociations, before it can reach Earth's surface. The less intense ultraviolet-B radiation (green), ranging between about 1070 and 950 terahertz, is almost completely absorbed, but a small portion of this bandwidth of radiation does reach Earth's surface, the actual amount depending on the concentration of ozone present in the stratosphere.

Ultraviolet-A radiation, on the other hand, of frequencies below 950 terahertz, plays little or no part in photodissociations, and therefore most UV-A penetrates to Earth's surface.

Snowball Earth: An Inconvenience

Sun was about 15 percent dimmer 2.5 billion years ago than it is today. Consequently, the exclusion from Earth's surface of Sun's most intense heat-producing radiation by ozone presented a huge problem at that time because it served further to cool the already cool planet. This was only made worse by the fact that Earth was gradually cooling due to the progressive decay of its original complement of radioactive elements. These two factors led to the onset of the first global glaciation that Earth had ever experienced, as evidenced by the very first appearance of glacial deposits and *glacial striations,* or parallel microgrooves, in rocks of that age. This so-called *Huronian* glaciation lasted for a very long time, from 2.4 to 2.1 billion years ago, or about 300,000,000 years, and probably covered the entire planet in ice.

The actual cause of this and other subsequent global glaciations during the *Precambrian eon* (the time before 541 million years ago) is hotly debated, however. The above interpretation makes sense in that no known evidence appears to contradict it, and the appearance of oxygen in the atmosphere coincides in time with the rock record of the first global glaciation. In contrast, some have suggested the rather contrived idea that prior to oxygen entering the atmosphere, Earth had a lot of methane, a greenhouse gas, in its atmosphere (why is unspecified), which reacted with oxygen at the time, and was thus removed, but I know of no evidence to support this theory. A similar general cooling effect, however, would have occurred if an ozone shield had formed in the newly oxygen-rich atmosphere, and if this had served to exclude solar ultraviolet radiation from Earth's surface.

In addition to this, the waste oxygen produced by the cyanobacteria was toxic to them, and so the combination of the glaciation and the presence of toxic oxygen would likely have caused a massive die-off of life at that time. Despite this, the oxygen already released to the atmosphere by the cyanobacteria would have tended to stay there because with the planet enshrouded in ice, there would have been little with which the oxygen could react. This could help explain the long duration of "snowball Earth" conditions. Meanwhile, a few life forms managed to survive in the relatively oxygen-free environment beneath the ice.

In time, some life forms, presumably those that lived in the more oxygenated and better lighted regions of thinner ice, evolved resistance to the toxicity of oxygen. Meanwhile, some experimentation was going on in which smaller bacteria invaded larger ones and became functional parts of them. This caused the larger cells to sequester their genetic material within protective, membrane-enclosed nuclei.

This process, known as *endobiosis*, eventually led to a new evolutionary development, the *eukaryotic cell*. Unlike its precursor, the *prokaryotic cell,* typical of the ancient cyanobacteria and their relatives, the new cell type evolved an ability to colonize a variety of changeable environments.

It also developed a new kind of reproductive system that would allow it to survive in such changeable environments. This we know as sexual reproduction, and its special advantage is that it allows genetic differences to express themselves among successive generations. To organisms colonizing new and changeable environments, this was a considerable advantage over simple division, in which new generations are genetically identical with older ones, as was the case with the ancient cyanobacteria, which lived in stable shoreline environments and therefore didn't need such genetic flexibility.

Under these frustrating and chilly conditions, life on Earth developed rather slowly, a period that has been called by some the "boring billion." During this time, Sun's radiative output gradually increased, and as Earth's rocky substance cooled, due to the gradual reduction in radiogenic heat within the planet, plate tectonics went from a "hot subduction" phase through a transition to the "cold subduction" phase that is in effect today.

The Resurgence of Life on Earth

By about 750 million years ago, the breakup of the supercontinent Rodinia began, with extensive rifting and consequent volcanism. This would have introduced substantial volcanic chlorine into the stratosphere, causing ozone thinning and consequent warming.

This effect, together with a slightly warmer Sun, again produced an environment that was favorable to the development of life. The renaissance came at a price, however, since rejuvenated photosynthetic life produced correspondingly more oxygen, and that increased the production of ozone in the stratosphere. By about 715 million years ago, Earth again slipped into a global glaciation, known as the *Sturtian*, which lasted almost 60 million years. It was followed by an interval of 10 million years without evidence of glaciation, which was in turn followed by another global glaciation, the *Marinoan*, which lasted about 25 million years.

Evidently, by about 575 million years ago, the conditions that would favor the proliferation of life on Earth took effect. Probably the most important of these was the gradual raising of the pH, or alkalinity, of the oceans to the point at which they could sustain multicellular life forms. The single-celled life forms that had dominated the planet throughout the Precambrian eon could thrive under a wide range of pH conditions with minima as low as 0.6, but most modern multicellular organisms require a pH range between 6.0 and 9.0 to thrive.

Presumably, the world's oceans would have had a pH well below 6.0 for most of Earth history, and thus would have been unfavorable for the maintenance of multicellular life, but why? A clue exists in the composition of the atmosphere, which is currently dominated by molecular nitrogen. The principal source for this gas is the denitrifying bacteria that dwell in the anaerobic (oxygen-free) environment of the sea floor. There are several species of denitrifiers, all of which utilize nitrate ion dissolved in seawater to metabolize carbon-based materials, but an overall reaction can be written as follows:

$$2\ NO_3^- + 10\ e^- + 12\ H^+ \rightarrow N_2 + 6\ H_2O$$

(two nitrate ions plus 10 electrons plus 12 protons equal one molecule of diatomic nitrogen plus 6 water)

Notice that in this reaction, hydrogen ions (H^+) are removed from seawater. Since pH is defined as the negative logarithm of the hydrogen ion concentration, it can be seen that the metabolic action of the sea floor denitrifiers utilizing nitrate ion in their metabolism gradually raised the pH of seawater. In the early Precambrian, nitrate was a fairly abundant component of seawater, but by the beginning of the Phanerozoic, it had been bacterially reduced to only trace amounts.

By this time, dissolved oxygen had also spread throughout the full depth of the ocean, driving the anoxic (oxygen-free) realm and its primitive microbial inhabitants to below the water/sediment interface on the sea floor. Also contributing to this new, life-favorable world was a warming Sun and continental rifting that produced voluminous, basaltic eruptions that emitted enough chlorine to make it finally warm enough on Earth for oxygen to rise to its present level in the oceans and atmosphere without triggering global ice conditions on Earth.

The thing that actually made it all possible, however, was something that had been obvious to us for centuries, but nevertheless, we failed to realize its true significance. For all of prior geologic time, the vast majority of life on our planet had been photosynthetic, and therefore, as it proliferated, it produced an abundance of the waste product, oxygen. Since about 575 million years ago, however, in the late Ediacaran and the following Cambrian, Ordovician, and Silurian periods, a new form of life became dominant. That new form was animals, which rely on *heterotrophy*, or the consumption of other organisms, for their supply of carbon, as opposed to *autotrophy*, or the provision of its carbon supply through photosynthesis.

In consequence of this new lifestyle, animals produced no waste oxygen, and therefore they couldn't strengthen the ozone layer and thus reduce the influx of solar radiation received by Earth's surface. This meant that the proliferation of this new life form couldn't bring on global glaciation, as prior proliferations of photosynthetic life forms had done. Since these animals preyed both on each other and on Earth's existing autotrophic forms, they effectively kept the latter in check, thus further restricting life's overall output of oxygen. In time, however, as they became larger and more complex, oxygen became more important to their more active lifestyle, and so a balance was eventually struck.

The Most Probable Cause of Global Warming

With this essential background, we can now consider the actual mechanism of volcanic ozone depletion that is the reason for this chapter and for the inclusion of ozone depletion in my graph. The underlying idea came from Peter Ward's work in 2006 with the GISP2 ice cores from Summit Camp, Greenland. He noticed that every time there was an abrupt rise in temperature, as recorded by the $\delta^{18}O$ *proxy* for temperature in the cores, there was also a large presence of volcanic sulfate in the same ice layers from volcanic eruptions in nearby Iceland.

This relationship applied universally to a series of 25 sudden, pseudo-cyclic warmings, called *Dansgaard-Oeschger events*, followed by slower cooling, that occurred throughout the most recent glacial period, and also to the major warmings at the end of that period, including the great *Preboreal warming* that finally ended the ice age, 11,950 to 9,375 years ago. During this final grand episode of warming, the ice record in Greenland indicated an average of about one major basaltic eruption every four years.

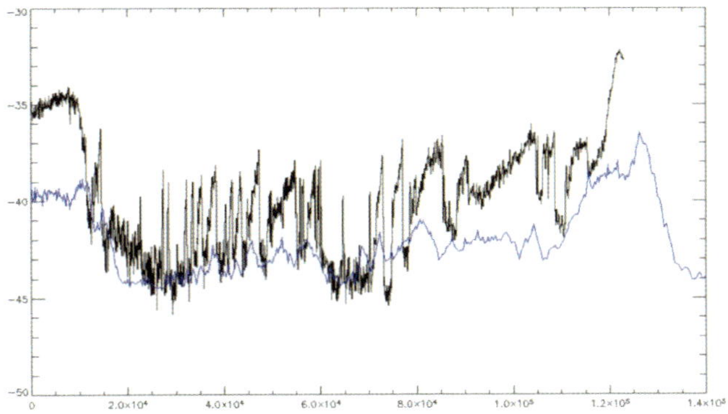

The figure above, which gets younger toward the left, shows all this in black in the form of the $\delta^{18}O$ oxygen isotope proxy record for temperature. The blue curve is a similar record from Antarctica for comparison, but this curve is considerably more muted and without either the dramatic jumps in temperature or the accompanying sulfate record because there were no nearby sources of volcanic activity, as there were in the Greenland cores.

Ward was puzzled by his discovery of volcanism associated with warming because it was already well-known by both the geological community and the general public that "volcanoes cause cooling." Such cooling happens because major *explosive* eruptions produce tall clouds that deliver vast quantities of volcanic ash, water vapor, and sulfur dioxide, as well as other substances, into the lower stratosphere, as is shown in the photograph below of the 1991 eruption of the *andesitic* volcano Mount Pinatubo in the Philippines.

Ash of a certain size, and sulfuric acid formed by the reaction of sulfur dioxide with oxygen and water vapor, are effective reflectors of sunlight, and their presence in a volcanic pall spread out in the stable stratosphere can cause cooling of the underlying troposphere for three years or more.

The crucial difference is, however, that the volcanoes in Iceland that produced the sulfate in the Greenland ice are *non-explosive,* and so they don't deliver eruption clouds of ash, sulfur dioxide, and water vapor to the lower stratosphere. Icelandic volcanoes, like those in the Hawai'ian Islands, are *basaltic*, and they produce a higher-temperature, more fluid type of lava that erupts quietly and is not viscous enough to cause explosions. In other words, it's only the explosive andesitic type of volcano that is responsible for global cooling. The non-explosive basaltic type, on the other hand, is evidently responsible for the precise opposite effect, global warming (see photo below of the 2014-15 eruption of Iceland's Bárðarbunga), but why?

All magmas (molten, subterranean rock), and the erupted lavas that form from them, contain gases, such as the aforementioned sulfur dioxide, which are released during eruption. The most common of these are water vapor, carbon dioxide, and sulfur dioxide. At first, Ward tried to establish a relationship between sulfur dioxide and warming, and he published a paper on that, but his results were less than convincing.

Lavas produce other gases in lesser amounts, however, including the hydrides of such *halogen* elements as chlorine and bromine, HCl and HBr. Ward turned his attention to these instead, because he recalled that the halogens

they contained had recently been shown to be agents of catalytic ozone destruction, because of which their production was banned worldwide by the *Montreal Protocol on Substances that Deplete the Ozone Layer*. He reasoned from this that if chlorine released to the atmosphere at ground level from anthropogenic spray cans and refrigerators could impact the ozone layer, as they have been shown to do, then certainly halogens from volcanic compounds released at ground level could do the same thing.

As it happened, the concern that led to the drafting and international ratification of the Montreal Protocol in the last two decades of the 20th century was not with global warming but with sunburn and genetic mutation, both of which are caused by ultraviolet-B radiation, which, it was immediately realized, would be able to reach Earth's surface in greater amounts if the ozone layer were thinned by chlorine-bearing anthropogenic compounds. Oddly enough, it apparently occurred to no one that this highly intense kind of solar radiation, with 48 times the frequency of the infrared radiation emitted by Earth and absorbed by carbon dioxide, was also capable of warming Earth, a conclusion that really should have been quite obvious from the great power of this radiation in causing sunburn. In fact, carbon dioxide plays no part in the absorption of UV-B, all the absorption of this radiation taking place in the water of the sea, the land surface, and in the body tissues of plants and animals.

Ward then researched the large igneous provinces (*LIPs*) throughout the Phanerozoic, or the last 541 million years of Earth History, and found the same relationship between sudden warming and massive outpourings of non-explosive, basalt lava, and he realized that this simple mechanism could therefore not only explain sudden warmings of the Earth system, but also the many great extinctions that usually accompanied these great outpourings of basalt. Conversely, frequent eruptions of explosive volcanoes could produce global cooling. Interestingly, whereas non-explosive volcanoes are characteristic of the inner, or *constructional*, edges of Earth's laterally spreading *tectonic plates*, the explosive kind is found at the outer, or *destructional*, edges of the plates where they sink beneath adjacent plates. Together, Ward reasoned, these two very different types of volcanic eruptions, both occurring simultaneously at different locations on Earth's tectonic plates, could account for the heating and the cooling of the planet.

Reactionary Rejection

Elated by this discovery, he then wrote up his findings and submitted them for publication in a peer-reviewed journal, but to his great surprise, not one of the seven publications to which he eventually submitted them would publish these results. Most rejected the paper without comment, but one telling remark revealed the reason: "This paper shows a total disregard of all known science." In other words, this reviewer, and probably the others as well, was evidently so indoctrinated by the conventional "wisdom" regarding matters of global warming that instead of welcoming a new, rational explanation for observed phenomena, he denied Ward's ideas outright.

I, however, at least recognized the value of Ward's important contribution, and therefore I offered to collaborate with him on a book that outlined the theory. That work, entitled "What Really Causes Global Warming? Greenhouse Gases or Ozone Depletion?" is now available on amazon.com by Peter Langdon Ward, edited, and with a foreword, by me. In the course of editing, I constructed the graph, discussed earlier, showing that carbon dioxide couldn't be the real cause of warming, but that ozone depletion could well be, and that graph is included in Ward's book.

Thus, there are at least two recently discovered pieces of evidence, both based on hard data from the Earth system, that now stand in contrast to the generally accepted but still only theoretical view that carbon dioxide is responsible for global warming: 1) Ward's discovery of the coincidence of sulfate from non-explosive volcanoes with warming, together with his consequent implication of ozone thinning from basaltic halogen releases allowing greater irradiation of Earth's surface by solar ultraviolet-B radiation, and 2) my graph showing that the maximum effect of carbon dioxide variability in the atmosphere, occurring in May, comes two months too late to affect warming anomalies, which peak in March, but the peak in the variability of ozone depletion also occurs in March, therefore allowing ozone depletion to be a possible, and indeed a likely, cause of global warming. In addition, there is the observation by infrared astronomers that back-radiation from carbon dioxide in wavelengths that correspond to Earth's surface temperatures does not interfere with their astronomical observations.

Why Earth is Still Slowly Warming

One final consideration is the persistence of record high global temperatures since the Montreal Protocol was ratified and the fact that temperatures seem to be increasing slowly, along with sea level, although not by any means at the rate that has been predicted by the computer models. How can Ward's ozone depletion model account for these trends, and for the related fact that worldwide, glacier and sea ice appears to be receding, except, paradoxically, in Antarctica?

The following graphic from the Australian government shows the probable reason for this. As noted, chlorine and bromine destroy ozone *catalytically*, that is, they are recycled during the ozone destruction process so that one chlorine atom can easily destroy 100,000 ozone molecules before it leaves the stratosphere, and chlorine and its precursors have relatively long residence times in the stratosphere, on the order of 50 to 300 years.

This means that the ozone destroying effects of chlorine compounds in the stratosphere are very persistent, and gradual warming, or at least the failure of global cooling to return to the planet, is likely to continue for some time to come, probably at least until the mid twenty-first century. Of course, mainstream thinking attributes this to rising carbon dioxide, but there are at least three problems with this. First, the increase in temperature is not steady, whereas the rise in the Keeling carbon dioxide base curve is. Second, the ozone depletion model fits

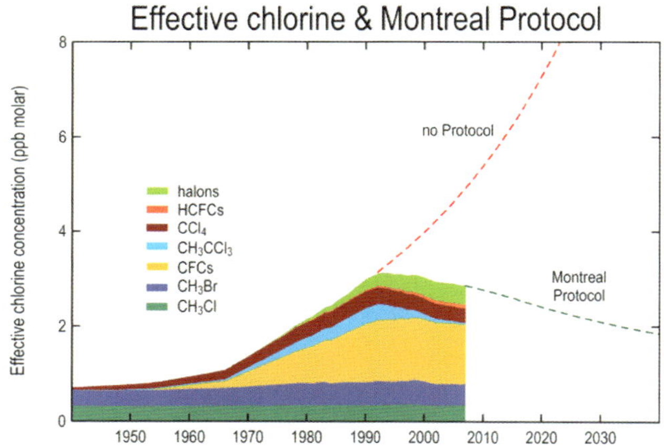

the actual temperature behavior of the atmosphere much better than the carbon dioxide model, and third, the computer models greatly overestimate the modest degree of warming that is actually taking place. All this argues for serious consideration of Ward's ozone depletion model.

3. Venus and Mars

A Pressing Matter

The dead planet Venus, shown above left without its dense atmosphere in this composite photograph of Venus, Earth, and Mars, is often held up as a poster child for "runaway" greenhouse warming. The average surface temperature of Venus is 462 °C (735K or 864 °F), the highest in the Solar System, surpassing even the hottest spot on Mercury's midday equator by 35 °C even though Mercury is almost twice as close to Sun as Venus. Whereas Mercury has almost no atmosphere, Venus has a thick atmosphere of almost pure carbon dioxide—96.5 percent—the remaining 3.5 percent being volcanic nitrogen.

The presence of this dense carbon dioxide atmosphere on super-hot Venus has been taken as prima facie evidence that a runaway greenhouse effect is at work on that planet, but in fact it shows nothing of the kind.

To understand this, there are a few other things on the planet Venus that must be taken into account, but that usually aren't considered in the rush to identify greenhouse conditions there. The first of these is atmospheric pressure at the surface, which is an astonishing 92 bars, or 92 times the pressure of Earth's surface atmosphere. There is a well-established relationship between the pressure and the temperature of gases, no matter what kind they are. That relationship is known as *Gay-Lussac's Law*, which states that:

$$P_1/T_1 = P_2/T_2$$

in which P is pressure in some chosen unit, and T is the absolute, or Kelvin, temperature. If P_1 is taken to be the pressure of Earth's atmosphere in bars; T_1 is Earth's temperature of 288 K (interestingly, Earth and Venus have about the same ideal temperature, largely because of Venus's higher *albedo*, or reflectivity, which causes it to absorb less of Sun's radiation than it otherwise would, even though it is much closer to Sun than Earth); and P_2 and T_2 are the equivalent values on Venus, then $T_2 = T_1P_2/P_1 = 288x92/1 = 26{,}496$ K, or 36 times Venus's actual surface temperature of 735K and almost five times the temperature of Sun's surface! Thus, the nearly hundredfold greater pressure alone in Venus's atmosphere should be sufficient to produce a surface temperature that is much greater than Earth's.

Using the Arrhenius formula, which specifies a rise of between 4 and 8 °C for each doubling of CO_2, then Venus's 96.5 percent CO_2 atmosphere should have a temperature equal to Earth's surface temperature of 288 K plus 18 doublings of 400 ppm (the present concentration of CO_2 in Earth's atmosphere) times (4 to 8K) = 360 to 432K.

So, by Gay Lussac's Law, the surface of Venus is far colder than it should be, and by the Arrhenius formula, it is three to four hundred K hotter than it should be. What can we tell from this? The answer: absolutely nothing. Looking first at the Gay-Lussac effect, the reason why Venus's surface temperature is way below what this law predicts it would be is that the atmosphere of Venus is not a confined system. Atmospheric gases are mobile and free to circulate, and as soon as Sun heats up the surface gas, the gas expands, becomes lighter, and rises to higher levels in the atmosphere, where the pressure is much lower, which cools the gas to a temperature far below what it would attain if it were locked in place at the surface. This buoyant rise is the start of the great winds that are known to whip across the face of Venus with constant ferocity.

Now, considering the Arrhenius formula, any possible effect due to greenhouse warming would be so small (a maximum of 144 °C) compared to the Gay-Lussac pressure effect that it would be undetectable in the rush of warming air away from the hot surface. Therefore, there really is no basis for the claim that Venus is a classic example of a "runaway greenhouse effect." Further, there is no reason to assume that greenhouse warming would be any more favored on Venus than on Earth. The same problem of there being no evident heat source independent of solar radiation should apply on Venus as on Earth.

Deuterium and the Importance of Moon

A few other interesting observations pertain to the second planet. First, Venus is bone dry except for a little water that has combined with volcanic sulfate to produce clouds of sulfuric acid in the atmosphere. Coupled with this is the fact that Venus has the highest deuterium to hydrogen ratio in the Solar System. *Deuterium*, or heavy hydrogen, is an isotope of hydrogen that has one neutron in its nucleus in addition to hydrogen's defining single proton. Hence it's about twice as heavy as normal hydrogen. On

Earth, there's about one deuterium atom for every 6,700 normal hydrogen atoms, but on Venus, there's one deuterium for every 62.5 hydrogens, thus increasing the ratio over a hundred fold, indicating that somehow, the planet has lost almost all of its original complement of normal hydrogen, leaving it much enriched in the heavier isotope, deuterium.

This makes sense when one considers several factors. First, it's very unlikely that Venus's anomalous ratio is original, and consequently the planet probably started out having the same very low ratio of deuterium to hydrogen that characterizes the Solar System in general. Second, the 50 percent lighter hydrogen is more likely to accelerate to Venus's *escape velocity,* or the speed necessary to escape the planet's gravity, than is deuterium, hence normal hydrogen is preferentially lost from the planet by this mechanism over time. Third, the only abundant source for hydrogen and deuterium on the various bodies of the Solar System is water.

It seems, then, that some mechanism has broken up water molecules on Venus over time, whereupon the lighter hydrogen preferentially escaped to space, leaving the oxygen to combine with such reducing surface materials as basalt, giving Venus's surface its reddish color and ultimately depriving the planet of its original complement of water, which could well have been similar to Earth's.

The reason for this loss, and for a similar, but less comprehensive loss of water from Mars, is debated among planetary scientists, but again, because a common habit of most scientists is to look at things one at a time rather than at the whole picture, a rather important consideration has simply been ignored. That is the fact that Earth is unique in the Solar System in having a large, natural satellite, Moon, which has a mass of 1/81, or about 1.2 percent, of Earth's. The other important consideration here is that Earth is not perfectly spherical, but its rotation causes it to bulge slightly at the equator, giving it the pumpkin-like shape of an *oblate spheroid* with an equatorial diameter that is about 42 kilometers longer than its polar diameter.

Moon's mass is sufficient to exert a gravitational pull on this equatorial bulge, which results in two things: first, it gives Moon's orbit a slight inclination, 5.14 degrees from the *plane of the ecliptic* within which most of the Solar System's planets and their satellites orbit Sun, and second, and most importantly for this discussion, it serves to stabilize Earth's rotational axis to within a range of two and a half degrees in total, its present value being 23.4 degrees from a perpendicular to the plane of the ecliptic. Evidently, this two and a half degree range hasn't changed significantly over geologic time. The same is not true of Venus and Mars, both of which lack a large, natural satellite like Moon. This is important because this lack allows the

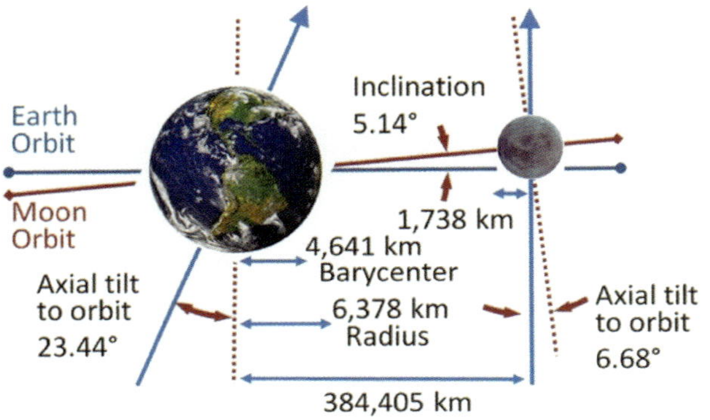

gravitational pulls of other planets to cause the rotational axes of both these planets gradually to swing freely from zero to 180 degrees and beyond. This flexibility has two important effects. First, depending on the degree of axial tilt, it creates dramatically changing and destabilizing conditions of seasonal insolation through the course of an orbit around Sun.

Second, any orientation that points these planets' polar regions in the general direction of Sun during either of the two *solstice* (summer) seasons exposes those regions to relatively intense solar irradiation, something that Earth doesn't experience with its modest axial inclination. This exposure would have provided an opportunity for greatly increased evaporation of any water located in the higher latitudes, followed by the photodissociation of evaporated water molecules by sunlight high in the atmosphere.

In the case of Venus, rotational axis inclination has gone to extremes, and the planet is actually upside-down, having tilted 176 degrees from a vertical to the plane of the ecliptic, and in consequence, it has a very slow retrograde (reverse) rotation of one turn in every 243 days. This means that any given point on the planet's surface is exposed to constant solar irradiation for that same period of time as Sun slowly transits the sunlit side of the planet. That the dark side of Venus has the same temperature as the light side is a reflection of the violent winds that constantly redistribute heat across and above the planet's surface.

Mars, on the other hand, is currently very similar to Earth in its axial tilt and its rotation, but at least the former of these is subject to change over time. Analysis of the deuterium to hydrogen ratio on Mars indicates that it is a little over five times that ratio on Earth, rather than a hundred, as it is on Venus, indicating a lesser amount of water loss, yet still enough to deprive the planet of any standing water on its surface.

Thus, because of the presence of Moon, Earth has been able both to retain its original complement of water and to develop life, whose various carbon-based processes require the medium of liquid water in which to operate. This fact should give some food for thought to exobiologists, who are perpetually hopeful to find evidence of life on other observable planets. It's quite clear that if they are to do so, any such planet must have a large, natural satellite that stabilizes its rotational axis as Moon does for Earth. On both Venus and Mars, which lack such massive satellites, the majority of these planets' carbon has ended up in their atmospheres.

4. Why Carbon?

We are Carbon Dioxide (Mostly)

Perhaps the best way to bring home the great importance of carbon to the Earth system is to reiterate the point made in the first chapter that everyone on Earth, including you and me, consists of a little more than two-thirds carbon dioxide, dry weight, and that if carbon alone is considered (i.e., removing the dioxide), we consist of eighteen and a half percent of it. In other words, you and I, along with all other living beings—plants, animals, mushrooms, or microbes—on Earth, are carbon-based organisms, and without carbon, we could not exist. If the Earth system had been devoid of carbon at its inception, life could not have evolved, and Earth would have been just another dead planet.

To any student of life on Earth, one of the first things he or she notices is the extraordinary diversity of organic compounds. There are over ten million of them that have been recognized, and many more that have not yet been. There are several reasons for this, but before looking at these, we should first take a brief look at the size and structure of the carbon atom.

First, carbon is a small atom. In its nucleus, it has six positively charged, heavy protons and usually six, but sometimes seven or eight, very slightly heavier, uncharged neutrons, which mainly serve to keep the positive protons from repelling each other and disrupting the nucleus. Rapidly orbiting this positively charged nucleus is a cloud of six very light, negatively charged electrons, which serves to balance the positive charge from the six nuclear protons.

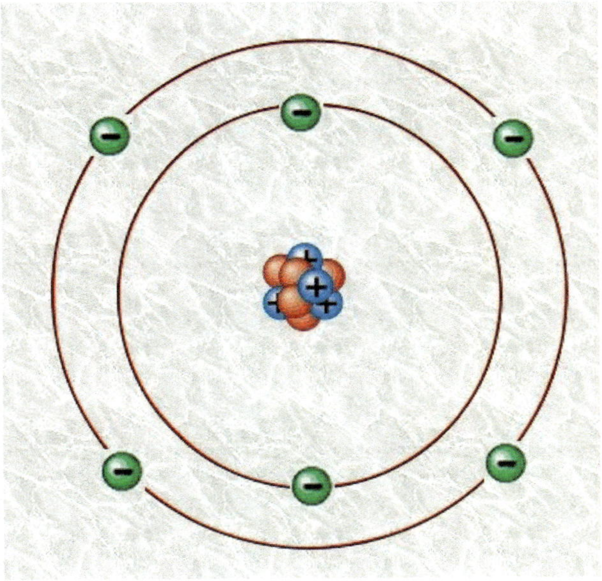

Two of these electrons are tightly bound around the nucleus (inner circle in the highly simplified diagram above), so they don't contribute to the process of bonding to other atoms to form carbon compounds. The other four, so-called *valence* electrons (outer circle in the diagram)

form bonds that tie the carbon atom to various neighboring atoms, including to other carbon atoms. Note that the circular orbits of these outer electrons are conventions only. In reality, the electrons whiz around the nucleus at dizzying speeds and with incredibly complicated, but definitely patterned motions.

When the valence electrons become involved in bonding to other atoms, this happens in two principal ways, depending on the configuration of the electrons. In one way, the bonds are all of the same strength, and they distribute themselves around the carbon atom in such a way that each one is as far from its neighbors as possible. This produces a tetrahedral pattern, or a regular polyhedron with four identical sides. When resting on one of these four sides, the tetrahedron looks like a three-sided pyramid, each side forming an equilateral triangle. The molecule methane, CH_4, which is of this type, in which one carbon atom is bonded to four hydrogen atoms, is shown below.

In the other style, the bonds are of two types, three of them equal and strong, and the fourth very weak, The three strong ones are equally distributed around the carbon atom, one forms a bond located perpendicularly to the plane of the triangle. In the example below, all the atoms shown are carbon, in this case, bonding to itself rather than to hydrogen or some other type of atom.

This difference in bonding styles gives rise to two very different structures in elemental carbon, one of them diamond, the hardest natural substance known ("a" in the second illustration below) because it is strongly bonded in all directions, and the other graphite, the softest of all natural substances ("b" in the illustration) because it consists of well *internally*-bonded sheets very weakly held to one another. In bonding to itself, carbon forms several other much less common structures, including lonsdaleite ("c"), found naturally only in meteorites, several kinds of fullerenes ("d, e, and f," named after eccentric architect Buckminster Fuller, who

used the structure "d" as the basis for some of his designs), amorphous carbon ("g"), and carbon nanotubes ("h").

Carbon is unique among elements in its strong ability to link to atoms of its own kind, and this is the basis for the vast array of carbon-based organic chemicals. Why does it have this unique ability? A look at the standard periodic table of the elements sheds light on this important question.

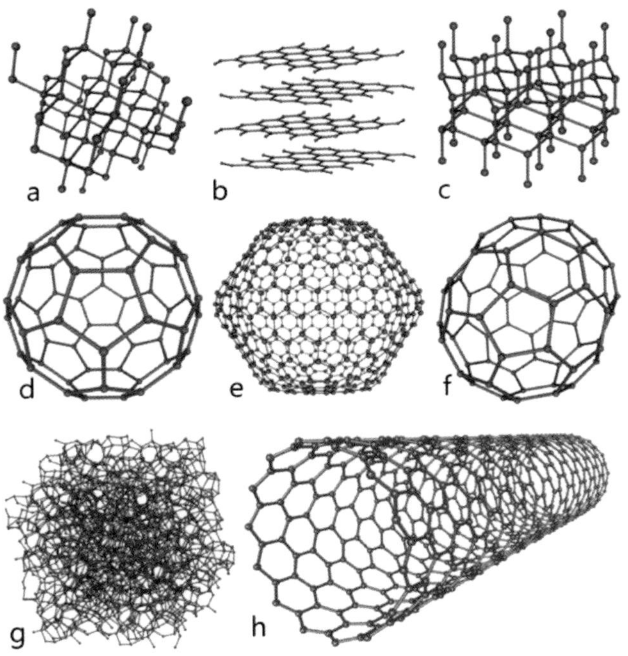

Carbon's Unique Place in the Periodic Table of Elements

Like all other natural things, the various atomic elements of which Nature is composed are well-ordered and very logical systems. This is well illustrated by the device by which atoms are classified, the latest edition of Russian chemist Dmitri Mendeleef's *periodic table*, the first version of which was published in 1869.

The table (shown below) is constructed on the basis of the number of protons in an element's nucleus, the order in which the clouds of orbiting electrons around atoms are filled, and chemical properties that recur periodically wih increasing atomic number. The first, which indicates the number of nuclear protons in the atoms, increases from left to right and top to bottom in the table, as indicated by the numbers in the individual boxes. The seven rows in the table correspond to the number of *shells* of electron clouds in the atoms, each larger than the one in the row above it. Each shell contains up to five *orbitals* of specific shapes within which the electrons travel. The eighteen columns of the table correspond to chemical properties, with metals (red), which have one or two valence electrons, on the left, and the *noble gases* (blue), with a full complement of eight valence electrons in their outer shells, on the right.

Group→ / ↓Period	1	2	3		4	5	6	7	8	9	10	11	12	13	14	15	16	17	18
1	1 H																		2 He
2	3 Li	4 Be												5 B	6 C	7 N	8 O	9 F	10 Ne
3	11 Na	12 Mg												13 Al	14 Si	15 P	16 S	17 Cl	18 Ar
4	19 K	20 Ca	21 Sc		22 Ti	23 V	24 Cr	25 Mn	26 Fe	27 Co	28 Ni	29 Cu	30 Zn	31 Ga	32 Ge	33 As	34 Se	35 Br	36 Kr
5	37 Rb	38 Sr	39 Y		40 Zr	41 Nb	42 Mo	43 Tc	44 Ru	45 Rh	46 Pd	47 Ag	48 Cd	49 In	50 Sn	51 Sb	52 Te	53 I	54 Xe
6	55 Cs	56 Ba	57 La	*	72 Hf	73 Ta	74 W	75 Re	76 Os	77 Ir	78 Pt	79 Au	80 Hg	81 Tl	82 Pb	83 Bi	84 Po	85 At	86 Rn
7	87 Fr	88 Ra	89 Ac	**	104 Rf	105 Db	106 Sg	107 Bh	108 Hs	109 Mt	110 Ds	111 Rg	112 Cn	113 Nh	114 Fl	115 Mc	116 Lv	117 Ts	118 Og

*	58 Ce	59 Pr	60 Nd	61 Pm	62 Sm	63 Eu	64 Gd	65 Tb	66 Dy	67 Ho	68 Er	69 Tm	70 Yb	71 Lu
**	90 Th	91 Pa	92 U	93 Np	94 Pu	95 Am	96 Cm	97 Bk	98 Cf	99 Es	100 Fm	101 Md	102 No	103 Lr

Notice that in this modern version of the table, the bicolored pink block of actinide and lanthanide elements at the bottom actually belongs within the space marked with three asterisks, but it is commonly placed below the table for convenience, as here.

On the table, carbon, C, occupies an isolated spot in bright green in the second row near the right side. Below it and to the right in rows three and four are three other elements in the same color category, phosphorus, P, sulfur, S, and selenium, Se. Bright green represents a category called *polyatomic nonmetals*, which takes into account two important tendencies: first, they share with all non-noble elements a tendency to add enough electrons from other atoms to fill up their outer, or valence, subshells to their full, stable capacity of eight electrons each. In this respect, they mimic the non-reactive, noble gases shown in light blue on the right edge of the table, each of which has a full eight electrons in its valence shell; second, these elements are unique in their tendency to *catenate*, or to link up with other atoms of the same species.

Carbon, in the second row, is a relatively small atom with respect to phosphorus, sulfur, and selenium. Because of this, the ratio of its electrical charge to its radius is especially high, and as a result, its tendency to catenate with other atoms of the same kind as itself is stronger than in the other three. Its small size, then, and its allotment of four valence electrons, are the primary reasons for carbon's unique ability to form strong chemical bonds with itself.

Boron, B, to the left of carbon in the table and in the gray *post-transitional metal* category, can also catenate to some extent, but it's easier for it to shed its three valence electrons, as a typical metal does, and thus be left with a single, completely filled first shell of two electrons, than it is for it to catenate. In nitrogen, oxygen, and fluorine, to the right of carbon, all of which are in the light green *diatomic nonmetal* category, it's easier for each of these elements to complete their second, outer, valence shells to a full complement of eight electrons by taking electrons from other, neighboring, usually *metallic elements* (red or beige, left side), leaving the latter with filled lower shells, than it is to catenate. When they do link up with other atoms of their same kind, they therefore limit such linkages to diatomic and sometimes triatomic states, as in O_2 and O_3.

Silicon, below carbon, is in the gray post-transition metal category because it is a larger atom, which makes it easier for it to shed the four electrons in its valence shell, as metals tend to do, than it is to catenate.

These, then, are the considerable benefits of carbon's unique position in the table, and they thus allow it to form strong catenation bonds to other carbon atoms more easily than to any other atoms. The fact that its outer, valence shell contains four of the electrons out of the stable, noble gas configuration of eight means that because of the equant distribution of the bonds made by these valence electrons, carbon can form chains of atoms that are straight, branched, or cyclic, and which can include single, double, or triple bonds and can include other kinds of atoms as well, as is illustrated in the drawing of the chlorophyll molecule below (note that for simplicity, the carbon atoms are left out rather than shown where two or more bonds intersect). All this gives the carbon atom extraordinary flexibility in the formation of organic compounds.

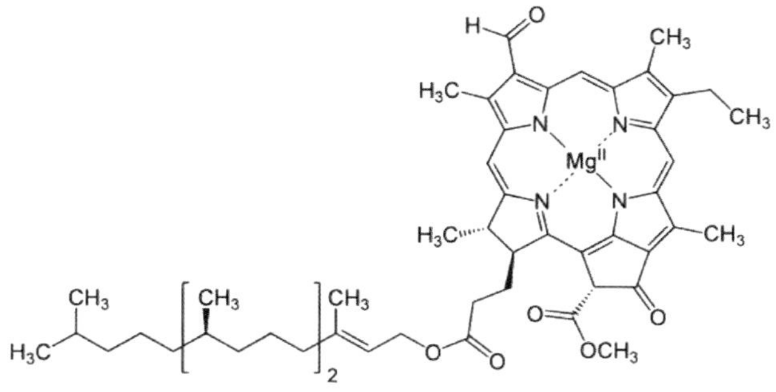

Carbon's Manifest Destiny

This special propensity of the carbon atom to make complex bonds with itself and with other atoms was by no means slow in being realized. As long as the "universal medium" of liquid water was abundantly available on Earth, and as long as energizing radiation kept pouring into the Earth system from Sun on a daily basis, organic chemicals of extraordinary complexity arose quickly and abundantly in the primordial soup that pervaded our young planet.

There was no lack of possible precursor molecules to the development of life in those waters, and it took only a very short time for the most suitable of these to assume the life-regulating roles that most of them still perform today. Thus, life arose very early in our planet's long history and has continued with a remarkable degree of consistency since then. It's quite true that complex, multicellular life had to wait for most of Earth history until conditions were right (i.e, a non-acidic ocean) for the *Cambrian explosion* of muticellular life about 540 million years ago, but the fact is that most of the mechanisms that are essential to these more advanced forms of life use exactly the same basic carbon chemistry that went into effect some four billion years ago with the first photosynthetic bacteria.

5. The Carbon Cycle

Where It All Goes

As with all components of the greater Earth system, carbon is involved in a complex of subcycles through various solid, liquid, and gaseous reservoirs. These cycles run on various scales, from the simple to the complex, and from the very fast to the very slow, and all of them are driven by the daily input of radiation from Sun. The reservoirs, of course, are the familiar ones of rock and soil, rivers and lakes, groundwater, the ocean, and the atmosphere, including that portion of the latter that pervades soils above the water table.

It came as something of a surprise to many students of the situation that the relative percentages of the dissolved substances in these various reservoirs are quite different and distinct, and of all the dissolved species, carbon is among the most variable. This can only mean that the processes that are responsible for establishing and maintaining the chemical compositions of the various reservoirs, and the rates with which they replenish themselves, are also quite different and distinct. A general principle is that the more a chemical species is reduced in concentration in moving from one reservoir to another, the more reactive and therefore important that species is in the cycling processes, so the finding that carbon in the ocean is a chemical species that is greatly reduced from river water is quite significant.

The table below indicates some of the percentage differences found in the composition of dissolved species in average river water and in ocean water arranged by reactivity and by decreasing percent in rivers.

Table 1. Chemical Species in River Water and Sea Water

Species	Rivers	Sea	Yrs in Sea
Reactive			
Bicarbonate (HCO_3^-)	31.90	0.42	110,000 (CO_3^{-2})
Calcium (Ca^{2+})	16.62	1.19	1,000,000
Silica (SiO_2)	14.51	-	20,000
Nitrate (NO_3^-)	1.11	-	short
Iron (Fe^{3+} colloid)	0.74	-	200
Conservative			
Sulfate (SO_4^{2-})	12.41	7.67	11,000,000
Magnesium (Mg^{2+})	4.54	3.72	13,000,000
Potassium (K^+)	2.55	1.11	12,000,000
Non-reactive			
Sodium (Na^+)	6.98	30.5	68,000,000
Chloride (Cl^-)	8.64	55.2	100,000,000

The chemical species in this table fall into three categories, the *reactive*, the *conservative*, and the *non-reactive*. The generally reactive ones include bicarbonate, calcium, silica, nitrate, and iron. Bicarbonate, for example, is by far the largest chemical component of river water by a margin of nearly two to one, but in seawater, it is reduced to a fraction of a percent, a reduction of approximately 76 fold! Calcium, river water's second most abundant ion, likewise suffers a decrease, in this case, one that is about 14 fold. Silica, the third most abundant component of river water, is virtually undetectable in seawater. Nitrate and iron, although minor constituents of river water, are, like silica, undetectable in seawater, but for very different reasons, as I will soon explain.

The non-reactive components, on the other hand, are sodium and chloride. Both of these are greatly concentrated in seawater over their concentrations in river water because neither is particularly important in life processes. We refer to the "salt sea" because it is so dominated by the non-reactive sodium and chloride ions. Conservative ions, which neither increase nor decrease very much as they enter the sea, include sulfate, magnesium, and potassium.

Why these changes? Various things happen to these river water ions as they enter the very different environment of the sea. For example, ferric iron, Fe^{3+}, a weathering product from various igneous rocks, combines with organic colloids in river water, but these negatively charged colloids are *flocculated*, or clumped, by positive ions like sodium in seawater, causing virtually all the iron to precipitate out in the estuarine environment before it can enter the sea. This precipitated iron is the main source for the frequently occurring iron-rich, green, mica-like mineral *glauconite* (or "greensand") in nearshore environments.

In the case of the other reactive ions, all of them have important roles in marine life processes and they are quickly taken up by these. *Bicarbonate,* HCO_3^-, and nitrate (NO_3^{2-}) are required to make protoplasm. Bicarbonate and calcium or silica are used to make shells. Nitrate is used instead of the more efficient oxygen by some anaerobic bacteria to oxidize protoplasm within oxygen-free bottom mud. Not shown here is phosphate, HPO_4^{2-}, another highly reactive ion, which is essential for making protoplasm.

The conservative ions, sulfate, magnesium, and potassium, have relatively minor, though still important, roles in life processes, so the rate of use of them is roughly commensurate with their rate of supply from the land. In the case of potassium, a natural inorganic sink, or avenue of removal from seawater, also exists in the form of clay deposits on the sea floor, which combine with potassium slowly to form a common clay mineral known as *illite* (after Illinois, which at one time lay under the sea).

What about the two non-reactive ions, sodium and chloride? One would think that they would actually accumulate slowly over time because of their general unimportance to living systems, making the sea ever saltier. This doesn't happen, however. Aside from their very minor use by biological systems, a plausible mechanism for their gradual removal is in Earth's inorganic plate tectonic system, which subducts into the mantle approximately the same amount of these two ions each year that is supplied by rivers.

Why does carbonate have such a high concentration in river water? Where does it come from in the first place? A first guess might be that it comes from the rock underlying the transporting rivers, but in that case, there would be practically no bicarbonate ion in regions with non-carbonate bedrock, which is not the case. Our attention, then, must turn to the plants that grow on the land surface. These plants aggressively pull carbon dioxide from the surrounding atmosphere, and use it to build their biomass. In fact, the dramatic reduction in atmospheric carbon dioxide that occurred in the Devonian period about 400 million years ago probably resulted when plants first became established on land.

Plants respire some of their biomass, releasing carbon dioxide, largely through their root systems. Plant roots also exude various organic substances, and these are oxidized by soil microorganisms as a source of energy. Since these root exudates are carbon-based substances, they, too, contribute to the addition of carbon dioxide to the soil atmosphere and to the groundwater, which ultimately feeds the flow of rivers to the sea.

The chemical changes involved in the conversion of carbon dioxide to bicarbonate ion in water are simple and straightforward. The reaction of CO_2 with water first yields carbonic acid, H_2CO_3, which then breaks down into hydrogen and bicarbonate ions, H^+ and HCO_3^-:

$$CO_2 + H_2O \rightarrow H_2CO_3 \rightarrow H^+ + HCO_3^-$$

(carbon dioxide plus water yields carbonic acid, which, in turn, yields hydrogen ion and bicarbonate ion)

This is an *acidifying reaction*, as is shown by the presence of hydrogen ion on the right side of the equation (since the *pH*, or acidity, of water is measured as the *negative* logarithm of the hydrogen ion concentration in the water, lower pH numbers are more acid), so the addition of carbon dioxide to the soil environment in this way renders it mildly acidic.

In the oceans, where photosynthesis removes hydrogen ion aggressively from the water, thereby rendering its pH higher than in soils, the above reaction is extended to produce two hydrogen ions and a carbonate ion:

$$CO_2 + H_2O \rightarrow H_2CO_3 \rightarrow H^+ + HCO_3^- \rightarrow 2H^+ + CO_3^{2-}$$

(carbon dioxide plus water yields carbonic acid, which yields hydrogen ion and bicarbonate ion, which yields two hydrogen ions and carbonate ion)

In this series of reactions, the addition of acid (hydrogen ion, H^+) on the right (higher pH) drives the series to the left, favoring the non-acidic species (e.g., CO2), whereas the addition of non-acidic species, on the left drives it to the right, favoring H^+. This has been called the *oceanic carbonate buffer system*, and it serves to stabilize the pH of the oceans against acidification from acidic sources. (Note that CO_2 is *not* by itself an acidic species, but that H+ is).

A rather minor factor in soil is that some disintegration of the underlying bedrock is always occurring, and the mildly acidic conditions contribute to this. A general equation that describes this rather slow process, using *anorthite*, or calcium feldspar, as a representative bedrock mineral is:

$$CaAl_2Si_2O_8 + H_2O + 2H^+ + 2HCO_3^- \rightarrow Al_2Si_2O_5(OH)_4 + CaCO_3 + CH_2O + O_2\uparrow$$

(Anorthite feldspar plus water plus two hydrogen ions plus two bicarbonate ions yields kaolinite clay plus calcium carbonate plus carbohydrate plus oxygen gas)

Unlike the previous reaction, there is quite a lot going on in this one. The left side of it shows four *inorganic* chemical species that interact in the weathering process, and the right side shows four products that result from the delivery of the weathered products to the sea. The first of these, *kaolinite clay*, $Al_2Si_2O_5(OH)_4$, is purely inorganic, resulting from the breakdown of feldspar during chemical weathering, and it accumulates in layers on the sea floor. In the second product, calcium carbonate, $CaCO_3$, also basically an inorganic substance, calcium from the weathering of feldspar has combined with bicarbonate ion in river water, but this product is extensively involved with organic processes, which cause calcium carbonate mainly to be deposited as skeletal material in the form of shells and internal "crystallites" within the bodies of living organisms. The third product, carbohydrate, or protoplasm, CH_2O, is exclusively organic, and it represents the operation of photosynthesis, which uses red and blue sunlight to strip hydrogen ions from water and adds them to carbon dioxide. Finally, this process yields a waste product of photosynthesis, oxygen, O_2, which is released to the environment, as is indicated by the upward-pointing arrow.

The Fundamental Equations of Life and Death

If we now focus on just the organic components of this equation and remove the weathering of anorthite from consideration except for its contribution of calcium ion, Ca^{2+}, (that is, if we remove $Al_2Si_2O_8^{2-} + H_2O + 2H^+ \rightarrow Al_2Si_2O_5(OH)_4$), the equation simplifies to:

$$Ca^{2+} + 2HCO_3^- \rightarrow CaCO_3 + CH_2O + O_2\uparrow$$

(Calcium ion plus two bicarbonate ions yield calcium carbonate plus carbohydrate plus oxygen gas)

In other books, I have called this *The Fundamental Equation of Life* because it combines two strictly non-organic ions into three products of life processes. One of these, carbohydrate, CH_2O, is strictly biological; another, calcium carbonate, $CaCO_3$, is an inorganic product of life processes; and a third, oxygen gas, O_2, is mainly a waste product of the life process of photosynthesis.

To be strictly complete, the Fundamental Equation of Life should actually be written with a hydrogen ion on each side, because the processes that are involved generate this ion on one side and consume it on the other, but these effectively cancel and thus leave the simpler form of the equation. Putting in these two hydrogen ions, we have:

$Ca^{2+} + 2HCO_3^{-} + H^{+} \rightarrow CaCO_3 + CH_2O + H^{+} + O_2\uparrow$
(Calcium ion plus two bicarbonate ions plus hydrogen ion yield calcium carbonate plus carbohydrate plus hydrogen ion plus oxygen gas)

The great importance of this lies in the fact that this is really two combined equations, *photosynthesis* and *carbonate secretion*, the former *deacidifying* because it consumes hydrogen ion (left side) and the latter acidifying because it generates that same ion on the right:

$HCO_3^{-} + H^{+} \rightarrow CH_2O + O_2\uparrow$ (photosynthesis)
(Bicarbonate ion plus hydrogen ion yield carbohydrate plus oxygen gas)

and

$Ca^{2+} + HCO_3^{-} \rightarrow CaCO_3 + H^{+}$ (carbonate secretion)
(Calcium ion plus bicarbonate ion yield calcium carbonate plus hydrogen ion)

What this means is that if the pH neutrality of the sea is to be maintained, these two equations *must run at the same rate*. In other words, if pH is to remain constant, then for every carbon atom from bicarbonate ion that goes into protoplasm, one carbon atom from bicarbonate ion must go into calcium carbonate. Only in this way are the hydrogen ions on opposite sides of the two equations able to cancel each other out, thus maintaining pH neutrality.

Here, I might mention that the carbonate secretion reaction is commonly written as

$$Ca^{2+} + 2HCO_3^- \rightarrow CaCO_3 + H_2O + CO_{2\,(aq)}$$

(Calcium ion plus two bicarbonate ion yields calcium carbonate plus water plus dissolved carbon dioxide)

which is a neutral reaction rather than an acidifying one, but as written, the left side of the equation implies that the one calcium ion will react with two bicarbonate ions. If this were actually to happen, the product of the reaction would be calcium bicarbonate, $Ca(HCO_3)_2$, which is a *soluble species*, and thus should be written as $Ca^{2+} + 2HCO_3^-$, but this is identical with the left side of the equation, which means that *no reaction actually takes place*. In reality, therefore, the single calcium ion only reacts with one of the bicarbonate ions, yielding calcium carbonate according to the carbonate secretion reaction given above. If this reaction is now subtracted from the equation under consideration, this results in the deacidifying reaction:

$$HCO_3^- + H^+ \rightarrow H_2O + CO_{2(aq)}$$

(bicarbonate ion plus hydrogen ion yield water and dissolved carbon dioxide)

which is simply a portion of the deacidifying carbonate buffer system, mentioned above, running in reverse. As the equation indicates, this would take the hydrogen ion produced in the secretion reaction and combine it with the second bicarbonate ion to form water and dissolved carbon dioxide, but this deacidifying function is actually performed by the photosynthesis conducted by the carbonate secreting organisms, thus rendering this buffering reaction unnecessary. In other words, only one of the two bicarbonate ions in the combined equation actually reacts with the single calcium ion to form insoluble calcium carbonate. The other bicarbonate ion is actually part of the carbonate buffer system, whose deacidifying function is rendered unnecessary by the uptake of hydrogen ion in the corals' photosynthesis reaction.

Why, then, is the pH of the oceans mildly alkaline, at a pH of about 8.2, and not strictly pH neutral, which is 7.0? The answer lies in the following graph, which shows that in order for the acidifying calcium carbonate secretion to run strongly enough to balance the deacidifying effect of the photosynthesis reaction, the pH of seawater must be maintained at about 8.2. This is easily done by the slightly more aggressive running of the photosynthesis reaction.

Also operating in the marine environment are the two other complimentary processes of *respiration* and *carbonate solution*, which, when combined, are simply the reverse of The Fundamental Equation of Life, and they are

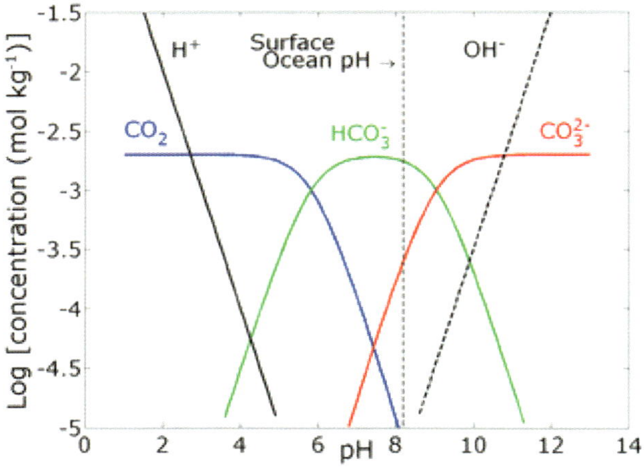

therefore expressed by the following reversed equation:

$$CaCO_3 + CH_2O + O_2 \rightarrow Ca^{2+} + 2HCO_3^-$$

(Calcium carbonate plus carbohydrate plus oxygen yield calcium ion and two bicarbonate ions)

Logically, therefore, I have called this *The Fundamental Equation of Death*, and as with the Fundamental Equation of Life, it, too, consists of two processes that produce hydrogen ions on either side of the overall equation that cancel each other when the two sub-equations are running at the same rate, over time. In general, this situation holds, but of course there are minor fluctuations one way or the other.

Interestingly, the Harvard geologist Heinrich Holland found, several decades ago, that over at least four billion years of geologic time, the chemical species in the oceans have never risen above twice their present concentrations nor have they fallen below half their present concentrations, and that the oceans have neither frozen solid nor risen above a maximum of about 40 °C, above which life processes find it increasingly difficult to operate. In short, he found that Earth's oceans have been remarkably constant over as long a period as yields us evidence of their existence, which is about eight ninths of all geologic time. The continued operation of these fundamental equations that I have just described over that long period and other mechanisms for the addition and removal of chemical species from the world's oceans are all responsible for this remarkable constancy.

The graphic below illustrates the interactive operations of the Fundamental Equations of Life and Death (which are written vertically to the left and right sides of the diagram, respectively), and how the various ions involved in the equations move about among the various compartments of the marine system within which they circulate.

In this illustration, boxes with sharp corners represent processes, those with rounded corners are chemical species (ions), and arrows are trans

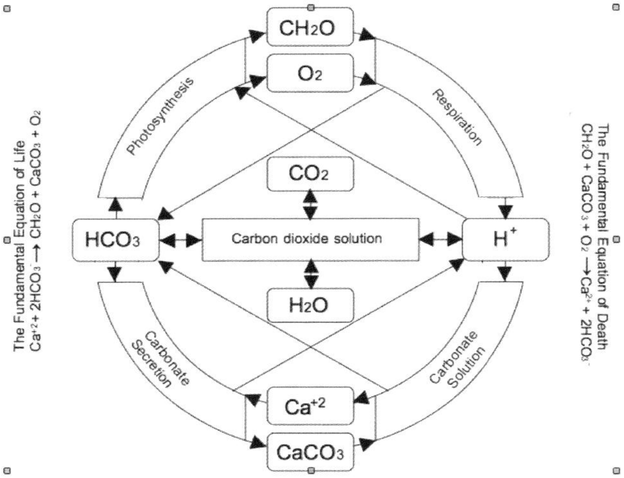

-fers due to the operation of the various processes involved. For example, the process of photosynthesis in the upper left draws in hydrogen ions, H^+, and bicarbonate ions, HCO_3^-, shown on the right and left of the diagram, respectively, and produces carbohydrate, CH_2O, and oxygen, O_2, both shown at the top. Similarly, the process of carbon dioxide solution and the carbonate buffer (center), which starts the whole complex of processes, draws in both carbon dioxide, CO_2, and water, H_2O, yielding bicarbonate, HCO_3^-, and hydrogen ion, H^+, as products. That these buffer processes are reversible as carbon dioxide *exsolves*, or escapes, from a warming ocean is indicated by the double-ended arrows governing this particular process.

Again, when all the processes shown in this diagram operate at approximately the same rate, the acidity of the oceans remains unchanged. Overall, this is the case, but of course there are local exceptions in both space and time.

Changes in Carbon Dioxide and Temperature Over Time

This is just one example of how the living Earth system has provided mechanisms for itself that assure the uninterrupted maintenance of conditions on the planet that are favorable to the continuation of life. If massive positive or negative excursions of carbon dioxide could have presented a problem of some sort, such as causing serious global warming, over geologic time, then certainly the Earth system (alias GAIA) would have responded to this putative threat with some sort of equivalent mechanism designed to prevent or compensate for such massive excursions, but in fact, no such mechanism has ever been identified. Moreover, several studies reveal that such excursions have actually occurred, as the following graphic showing various measures of past carbon dioxide concentration illustrates (time becomes more recent from right to left). As noted above, the high concentration of carbon dioxide in the early Phanerozoic was likely due to the absence of land plants at that time.

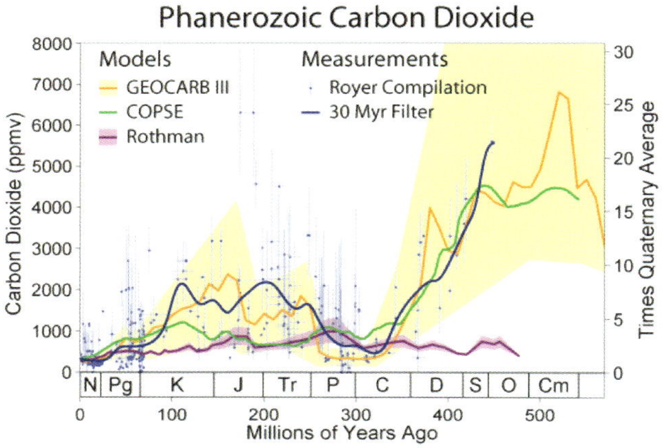

Although some attempts have been made to show that high values of carbon dioxide concentration coincide with hothouse conditions and low values coincide with cold periods, there is in fact no convincing correlation, and some episodes of global glaciation, such as the one in the late Ordovician period (O) actually coincide with high carbon dioxide values. In a September, 2017 article in "Climate" journal, W. Jackson Davis showed that correlation between CO_2 and proxies for temperature were either nonexistent or *slightly negative* for most of the Phanerozoic eon.

Many people have unfortunately been deceived by plots of paleotemperature and carbon dioxide variation from the Vostok Core in Antarctica, which show a tight correlation between these two variables, as in the graph shown below. Close examination, however, has revealed that although the two curves seem to rise rapidly together, the temperature curve, shown in blue, in general precedes the carbon dioxide curve, shown in green, by a matter of months. What is even more telling is that on the downward portion of the cycle the drop of the carbon dioxide curve lags behind the drop in the temperature curve on the order of centuries.

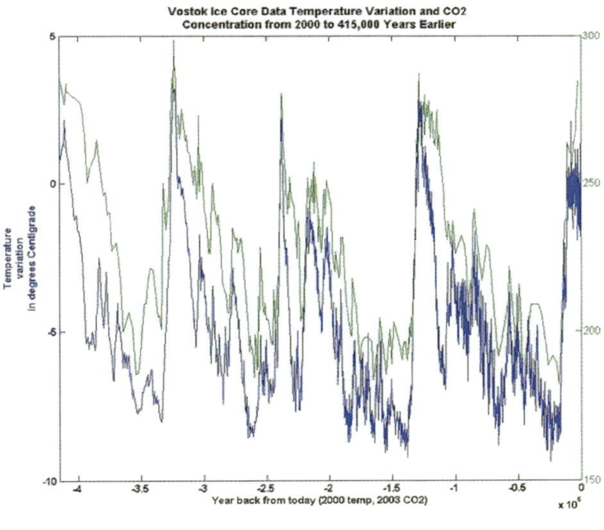

All this, of course, is a reflection of the known

fact that a cold ocean can hold more carbon dioxide gas in solution than a warm one can, and that as the ocean warms up, it abruptly releases carbon dioxide to the atmosphere. Conversely, as the ocean cools down again, it absorbs carbon dioxide from the atmosphere, but it does so much more slowly than it releases it on warming because the carbon dioxide is well dispersed throughout the atmosphere, so it takes much longer for it to reach the ocean surface and to be absorbed there.

In short, the Earth system has seen to it that it has mechanisms in place that guard against major changes that could disrupt the persistence of an environment that is favorable to the continuation of life. Naturally, there are cataclysmic events, such as meteoroid impacts and massive volcanic eruptions of great volume and long duration. These do have disruptive effects, and at least five massive extinctions of much or most life on the planet have resulted from just such events.

The Redfield Ratios

One example of the pervasive influence of carbon-based life on the planet Earth, as well as of its remarkable uniformity, is given by a relation called the *Redfield ratios* after the researcher who discovered the basis for it in the middle of the twentieth century. Redfield found that all the oceanic protoplasm that he tested had the same ratios of carbon to hydrogen, oxygen, nitrogen, and phosphorus. Those ratios stand in the proportions 106C:263H:110O:16N:1P. From this, Redfield developed an equation, which somewhat more accurately portrays the photosynthesis reaction:

$$106\ CO_2 + 122\ H_2O + 18\ H^+ + 16\ NO_3^- + HPO_4^{2-} \rightarrow C_{106}H_{263}O_{110}N_{16}P + 138\ O_2\uparrow$$

(106 carbon dioxide + 122 water + 18 hydrogen ion + 16 nitrate + hydrogen phosphate ion yields protoplasm + 138 oxygen gas)

Note that the main difference between the Redfield equation and the photosynthesis reaction is the addition in the former of relatively minor amounts of nitrogen and phosphorus, which are essential, though rather small, components of protoplasm. In Table 1, nitrogen is undetectable in seawater and phosphorus doesn't even appear, which indicates that both of these chemical species are *limiting nutrients* with extremely short residence times in the ocean, and that when they do appear in amounts sufficient to make protoplasm, they are immediately removed by blooms of life that suddenly appear to take full advantage of their presence.

From all the foregoing it is therefore quite clear that over geologic time the carbon cycle has come to dominate the reservoirs of life's essential nutrients and the various processes that distribute them throughout the Earth system. It is also clear that the Earth system itself has been modified by the operation of the carbon cycle in such a way as to create and maintain conditions that are most favorable to the existence of life on this planet. Despite occasional meteoroid, volcanic, and plate tectonic catastrophes, the Earth system has thus been maintained throughout geologic time in a state of highly ordered equilibrium favorable to the furtherance and preservation of life. In this respect, James Lovelock's vision of our planet as being itself a living organism is indeed compelling.

6. Beyond Doomsday

The Global Consequences of Climate Hysteria

In the first two chapters of this book, I have shown that the panic over the expected warming of Earth's climate due to an increase in the concentration of atmospheric carbon dioxide is unwarranted for two reasons. First, there exists no concrete evidence that such a relation exists in the real world, the concept only having support from complex theory, computer modeling, and the virtually religious attitude toward it held by many dedicated climate scientists, political operatives, the press, and the general public; and second, there is Peter Ward's alternative explanation for global temperature increases, involving the thinning of Earth's ozone shield by both anthropogenic and volcanic sources of chlorine. Ward's explanation accords much better than greenhouse warming with the existing evidence and with the ongoing behavior of observed global temperature changes over the past half century.

If this lack of a causal connection between increases in atmospheric carbon dioxide content and warming is indeed the case, then clearly doing anything to mitigate anthropogenic carbon dioxide emissions in the future will bring us absolutely no benefit whatsoever. It will simply waste a great deal of money and effort on the wrong solution to the warming problem, and it will cause great hardship for less-advantaged people and nations that would have to convert their energy sources to various inefficient, unsightly, and environmentally destructive non-carbon-based alternatives. It would also continue to demonize carbon, which, as I show here, is by any standard a truly remarkable element that has made possible the development and maintenance of life on this planet, and as such, deserves our utmost respect and gratitude, and definitely doesn't deserve to be regarded as a scapegoat. We literally owe our very existence to carbon and its myriad compounds.

That this unfortunate demonization of carbon is rooted in various agenda is virtually beyond question. Many careers have been built on the basis of greenhouse warming theory, and a lot of grant money has been spent on developing the theory and the computer modeling that purports to show its global effect. Powerful proponents of globalism and world government are solidly behind the concept because it presents us with a fearful premise on which to take action against carbon-based fossil fuels, which have powered the development of our advanced civilizations worldwide for well over a century. Replacing these with far less efficient, far more costly, and far more environmentally destructive alternatives makes very little sense, except that it gives globalists an opportunity to impose world governance that can dictate energy policy, and by extension other policies, to former autonomous nation states.

This carbon-demonizing globalist agendum will die hard, as there are many who have dedicated their hopes and careers to it. Furthermore, unfortunately, it serves to support the pernicious attitude toward science in the world today that consensus about a theory can substitute for hard evidence that supports that theory. Consensus simply can't do this, as I have stated above. Any theory remains unproven and invalid if it has not been shown by hard evidence that it actually works in the way it says it does, and in the case of greenhouse warming this critical step hasn't yet happened. In fact Ångström's 1900 results and my own in 2015 clearly show that it doesn't happen and that it can't happen, respectively. The experience of infrared astronomers confirm these conclusions, and the low frequency band over which carbon dioxide absorbs radiation shows why this is the case. Still, dedicted "warmists" claim that they're right about this, and they will undoubtedly continue to do so, even though trends in the behavior of Earth's climate system don't support their contentions. Ingenious, face-saving reasons will, as usual, be divined as to why reality deviates from theory.

In addition, there is an active campaign in

effect worldwide to suppress alternative views on this subject, and such epithets as "climate skeptic" and "climate denier" have been used regularly to discredit and marginalize those who do hold different views, and even those who simply show a disinclination to accept the "traditional wisdom" without question. No doubt this book will receive a similar treatment of vilification and trivialization, but perhaps it can help at least some to understand that the situation with global warming might in fact be somewhat different from the way in which it has been portrayed.

Going Forward

Looking to the future, what should we do, then? First, given the somewhat unexpected results of the 2016 election, I would think that the rather extensive federal funding of greenhouse warming research would be significantly curtailed, and despite the outrage that this will certainly cause among professionals and non-professional adherents to the concept, it would go a long way toward quelling emphasis on the theory, which only raises people's fears that it does, in fact, constitute a serious problem.

Given a new awareness of ozone depletion as an alternative scenario to warming, research in this area should be encouraged in order to identify actual links that show a connection between the two rather than just the compelling coincidence of events and phenomena that currently serves as the de-facto basis for hypothesizing a connection. Meanwhile, care should be taken worldwide to uphold and strengthen the provisions of the Montreal Protocol on Substances that Deplete the Ozone Layer in order to minimize the accidental and purposeful release of anthropogenic chlorine- and bromine-bearing chemicals and other substances that are known to have a catalytically destructive effect on stratospheric ozone.

The fact that the action of such substances is catalytic, combined with their relatively long residence times in the stratosphere, means, of course, that if the connection I describe above is indeed valid, then the problem of warming from ozone depletion is likely to be with us for some time to come, several decades, at the very least. Naturally, greenhouse warming theorists will be quick to ascribe such effects to greenhouse gases instead, and given the prestige with which this attitude is regarded, replacing it with ozone depletion theory will require considerable effort in the form of research and education.

Some Neglected Considerations

Meanwhile, considerable study is currently under way to assess the beneficial effects that actually result from a more carbon-dioxide-rich atmosphere, among which are the significant greening of the planet, allowing plants not only to grow better but also to extend their ranges into regions, like Africa's Sahel, where they formerly did too poorly to be able to survive under a less CO_2-rich atmosphere.

Ocean acidification is yet another problem that has been almost universally attributed to an increase in atmospheric carbon dioxide, but I find it quite interesting that those who write journal articles on this subject universally ignore such strongly acidifying gaseous oxides as those of sulfur and nitrogen, that almost always accompany CO_2 out of smokestacks and tailpipes. It could well be that these other oxides, which don't have the same kind of buffering mechanisms against oceanic pH change as carbon dioxide, and are much stronger, have more of an acidifying effect on oceanic waters than CO_2 does, but this needs to be proven by appropriate research. One compelling fact is that acidification is more of a problem in nearshore environments than it is in the open ocean, and the nearshore environments coincide with outfall zones of pollution from effluent sources.

Finally, it seems clear that the call to end the world's dependence on carbon-based energy sources as quickly as possible in order to avoid certain climate Armageddon has been made without regard to the many significant consequences of this global action. The civilization that we have today has been built on the reality of relatively inexpensive and readily available fossil fuels that have a very high energy content per unit weight. All proposed alternatives are considerably less efficient in this regard, and yet they are somehow expected to take over from fossil fuels as viable substitutes for energy production. This of course means that such non-carbon-based energy sources must inevitably be produced in far greater proportion than the fossil fuels that they are expected to replace.

It should be clearly understood that each of these alternative energy sources comes with its own set of problems. Solar panels cover space, which can conflict with other uses, including natural habitat, although installation on roofs is one good solution. Sun angle and storage batteries are also factors to be considered. Hydroelectric power drowns waterways, along which riparian natural ecology is inevitably destroyed. Nuclear power comes with an obvious hazardous waste disposal problem. Wind generators are unsightly, noisy, not very efficient, and they kill large numbers of migratory birds. Some people point out that domestic cats are responsible for bird losses as well, but that doesn't justify the addition of another source. Biomass energy replaces large tracts of natural forest cover with fast-growing, harvestable trees.

One little-appreciated fact is that in New England, for example, between 65 percent (southern states) and 90 percent (northern states) of forest land is privately owned. If the heavy taxation of fossil fuels or their outright prohibition makes forest wood more economical to use, it won't take long for New England to become deforested again, as it has been in the past, and recent gains in wildlife habitat would be lost, not to mention the concomitant detriment to New England's iconic image as forested country.

Looming over this specter of fossil fuel replacement is its implications for the many developing nations of the world. The promise of cheap fossil fuels for rapid development of such countries would be greatly muted by a transition away from the use of these resources, and the alternatives would all be more expensive. This would entail large, continuing, annual contributions of funds from the developed nations, which would be involved in the business of making their own transitions to non-carbon-based energy sources. Finally, decarbonization would render worthless the potential buried carbon-based fuel resources of many of the world's nations.

At the same time, it is quite clear that existing fossil fuel resources can serve to power the operations of our civilization for at least two or three centuries to come. The United States, for example, has by far the richest deposits of oil shale in the world, and the affordable development of this energy abundance has now become possible through new advances in technology and efficiency. Marine methane clathrate reserves on the continental shelves dwarf even the world's original untapped deposits of hydrocarbons, promising energy resources for many centuries to come. Nonetheless, two considerations stand in the way of development of all these deposits. This book is mainly concerned with the first of these, that is, the unproven assumption that increases in atmospheric carbon dioxide cause global warming. If in actuality they don't, then in calling for a transition away from carbon-based energy, we are simply tilting at the wrong windmill and shooting ourselves in the foot on the basis of an invalid assumption.

The second consideration, of course, is the essentially uncontrolled growth of human population, which serves to exacerbate every aspect of every environmental problem that we face, including the eventual exhaustion of the fossil fuel and marine methane clathrate resources. This issue was championed by Dr. Paul Ehrlich, who wrote about it in his 1968 book "The Population Bomb." Unfortunately, Ehrlich, though a good researcher and scholar, was overly alarmist in his doomsday predictions, all of which he predicted far too early, and all of which were easily overcome with appropriate actions in the sphere of technology. Consequently, Ehrlich's premise was unfortunately quite thoroughly discounted and dismissed, and we now have a world population that is at least double what he thought would be practically possible.

Nonetheless, we do live on a planet whose natural systems did not evolve with a view to accommodating the needs of an expanding human population, and as we do expand, the interests of other, non-human populations are commensurately compromised, as are the natural resources on which we humans are naturally and unnaturally dependent, including fossil fuels and, particularly, phosphate rock, a non-renewable resource that is essential for agriculture. Clearly, it is imperative that we start paying attention to the need to stop human population growth and preferably to reduce it to a level that is compatible with the planet's ability to serve our human needs over the long term. As a benchmark, one estimate of a sustainable human population for Earth is the number the planet supported in 1950 without the benefit of the green (i.e., chemical) revolution. That number was about 2.5 billion people.

Some argue that a world government is necessary in order to impose the sort of restrictions worldwide that would be required in order to control human population. I feel, on the other hand, that this problem can be better handled by individual nation states as they increasingly become aware of population-related problems that arise within their borders and take measures to redress them, such as taxation for having more than one child.

Although it may seem cruel, the shutting off of economic aid to nations that refuse to consider population problems is a very effective way to induce such nations to institute practical population-stabilization programs, and the promise of resumption of aid can be an equally powerful incentive in ensuring that such programs remain in place.

In closing, I would simply make the observation that in our modern world, important decisions are being made more on the basis of strong, ideological stances and on the weight of popular and professional opinions rather than on the basis of sound evidence and practical realities. In our haste to "do something," *disamenities*, i.e., the negative consequences of seemingly positive initiatives, are not adequately taken into account. If more people, given sufficient exposure to actual evidence, could be induced to think for themselves, more progress on such important decisions could be made.

The End

Did you enjoy reading this book? If so, why not write a review for the book's product page?
amazon.com/gp/product/B01N7ZXTID

Glossary of Italicized Terms

Absolute humidity: the water content of air, normally expressed as grams per cubic meter

Acidifying reaction: a chemical change that releases hydrogen ion, H^+

Albedo: the extent to which a surface reflects radiation, where 0 = total absorption and 100 = total reflectivity

Andesitic volcano: a volcano that erupts lavas intermediate between high-silica rhyolite and low-silica **basalt**

Anorthite: the calcium end member of the feldspar series; $CaAl_2Si_2O_8$

Anthropogenic global warming (AGW): the supposed warming of Earth's surface environment by the absorption and downward-directed re-radiation of infrared emissions from Earth by polyatomic gases

Banded iron formation: a Precambrian marine sedimentary deposit resulting from the reaction of dissolved **ferrous iron**, Fe^{2+} with oxygen produced by photosynthetic organisms; consists of layers of **hematite**, Fe_2O_3, or **magnetite**, Fe_3O_4, or both, alternating with shale or chert

Basaltic volcano: a volcano that erupts low-silica **basalt** lava

Bicarbonate ion: a soluble, mildly basic ion, HCO_3^-

Black body temperature: the temperature of a body calculated from the frequency of its highest amplitude radiation, assuming that the body has an **albedo** of zero

Broad-spectrum radiation: a band of continuous wavelengths of radiation emitted by condensed phases (solids and liquids) resulting from the **conduction of energy** within them.

Cambrian explosion: the proliferation of advanced life forms in Earth's oceans that began 541 million years ago, after deacidification by denitrifying bacteria rendered them habitable.

Carbohydrate: any compound of carbon, hydrogen, and oxygen, usually in a 1:2:1 ratio. Other elements may be present.

Carbon pollution: environmental damage theoretically caused by **greenhouse warming**

Carbon tax: a tax levied on the use of carbon-containing compounds thought to cause **greenhouse warming**

Carbon footprint: a quantitative measure of a person's use of materials thought to be capable of causing **greenhouse warming**

Carbonate secretion: the **acidifying** formation of skeletal material by marine organisms by combining calcium ion, Ca^{2+}, with **bicarbonate ion**, HCO_3^-, yielding calcium carbonate, $CaCO_3$, and hydrogen ion, H^+, the reverse of **carbonate solution**

Carbonate solution: the **deacidifying** reaction of calcium carbonate, $CaCO_3$, with hydrogen ion, H^+, yielding calcium ion, $Ca^{2+}$and **bicarbonate ion**, HCO_3^-, the reverse of **carbonate secretion**

Catenation: the tendency of an atom, especially carbon, to combine with other atoms of its own kind in chains, networks, or 3-dimensional structures

Causation: a condition in which one phenomenon causes another

Climate change: any process by which different environmental parameters become established in a given location; current usage implies human agency for this

Conduction of energy: the blending together of spectral lines of different wavelengths due to energy exchanges among molecules in mutual contact in solids and liquids

Conservative species: a chemical species in river

water that undergoes relatively little change in concentration upon entering the sea

Constructional plate edge: the inner, or trailing, edge of a spreading **tectonic plate**, marked by non-explosive **basaltic volcanism**

Correlation: condition in which two variables undergo comparable positive or negative changes at the same time

Covalent bond: a strong pairing of one to four **valence** electrons with **valence** electrons of another atom, or other atoms.

Dansgaard-Oeschger events: a series of 25 sudden warming events recorded in Greenland ice cores during the last glaciation

Deacidifying reaction: a chemical reaction in sea water in which hydrogen ion, H^+, is consumed

$\delta^{18}O$ proxy: a measure of paleotemperature based on the ratio of the two isotopes ^{18}O and ^{16}O

Destructional plate edge: the outer, or leading edge of a spreading **tectonic plate**, marked by subduction and explosive, **andesitic** volcanism

Deuterium: a heavy isotope of hydrogen containing one proton and one neutron in the nucleus

Diatomic: characterized by pairs of atoms covalently bonded to one another

Diatomic nonmetal: category of elements that tend not to **catenate**, but to fill their **valence** shells with electrons donated by **metals**

Disamenity: an unforeseen negative consequence of an action

Endobiosis: the formation of a **eukaryotic cell** through permanent invasion of a large **prokaryotic cell** by smaller **prokaryotes**

Escape velocity: the upward speed required to free an object from a planet's gravitational field

Eukaryotic cell: a larger and more modern cell type having linear chromosomes enclosed within a nucleus with a membrane, and several organelles also enclosed by membranes; the product of **endobiosis**

Exsolve: to come out of solution

Ferric iron: iron that has lost 3 **valence** electrons, Fe^{3+}

Ferrous iron: iron that has lost 2 **valence** electrons, Fe^{2+}

Flocculate: to form clumps that settle out of solution

Frequency: the number of times per second that a cyclical action is repeated; designated by the Greek letter ν (nu); the inverse of **wavelength**, λ; $\nu = 1/\lambda$

Gay-Lussac's law: increasing the pressure on a confined gas increases its temperature $P_1/T_1 = P_2/T_2$

Glacial striations: rigidly parallel micro-grooves in bedrock caused by rock debris embedded in overriding ice

Glauconite: greenish, iron-rich marine mica, $(K,Na)(Fe^{3+},Al,Mg)_2(Si,Al)_4O_{10}(OH)_2$

Greenhouse effect: see **greenhouse warming**

Greenhouse gas (GHG): a gas of triatomic or larger molecules capable of absorbing **infrared radiation**

Greenhouse warming: the theoretical but unproven heating of Earth by its own **infrared radiation** absorbed by atmospheric GHGs and re-radiated downward

Halogen: one of the **diatomic nonmetals** fluorine, Fl; chlorine, Cl; bromine, Br; and iodine, I

Hematite: the oxide of **ferric iron**, Fe_2O_3

Huronian glaciation: a severe, global glaciation that lasted 300,000,000 years, from 2.4 to 2.1 million years ago

Illite: a common, potassium-rich clay mineral, $(K,H_3O)(Al,Mg,Fe)_2(Si,Al)_4O_{10}(OH)_2,(H_2O)$

Infrared radiation: a **frequency** field having a bandwidth that ranges from 700 nanometers (428 **te-**

rahertz) to 1,000,000 nanometers (300 gigahertz)

Insolation: radiation from Sun, mostly visible light

IPCC: The United Nations Intergovernmental Panel on Climate Change in Geneva, Switzerland (parodied by some as the Intergovernmental Politicized Climate Cartel)

Kaolinite clay: $Al_2Si_2O_5(OH)_4$, an extreme chemical weathering product of feldspar

Lapse rate: the rate at which air temperature decreases with elevation in the **troposphere**

Limiting nutrient: an essential nutrient, such as nitrogen or phosphorus, that is locally in short enough supply to limit population growth

Line spectrum: a set of discrete wavelengths of radiation emitted by gases, whose molecules are not in mutual contact, and therefore don't exhibit **conduction**

LIP: a large igneous province, typically **basaltic**, and located at the proximal edges of tectonic plates, that accompanied occasional aggressive **plate tectonic** activity during geologic history

Magnetite: the sesquioxide of **ferrous** and **ferric iron**, Fe_3O_4, which is 2/3 Fe^{3+} and 1/3 Fe^{2+}

Marinoan glaciation: a "snowball Earth" event that was probably global in scope and lasted 15 million years, from 650 to 635 million years ago

Metallic element: an element having 1 or 2 **valence** electrons that readily releases these to other atoms, and does not **catenate**

Montreal Protocol on Substances that Deplete the Ozone Layer: an international treaty to protect the ozone layer, ratified from 1989 to 1998 by 196 nations and the European Union

Noble gas: helium, He; neon; Ne; argon, Ar; krypton, Kr; Xenon, Xe; and radon, Rn; all of which are nonreactive because their **valence** shells are full

Non-reactive species: a chemical species, including sodium, Na+, and chloride, Cl^-, which is relatively scarce in river water but of little use to organisms and hence relatively concentrated in sea water

Oblate spheroid: the shape of Earth, flattened at the poles and bulging at the equator due to the planet's spin

Orbital: a precisely defined space around an atom occupied by a definite number of rapidly circulating electrons; orbitals can be expanded, stepwise, by radiation of sufficient **frequencies**

Ozone: triatomic oxygen, O_3

Periodic table: an exceedingly useful graphical device invented by Dmitri Mendeleev whereby elements are arranged according to the number of protons in their nuclei, the addition of **valence** electrons, and chemical properties

pH: the negative logarithm of the concentration of hydrogen ions, H^+, in water. In pure water, at pH = 7, H^+ and hydroxyl, OH^-, are equal. As pH decreases, H^+ increases, and the reverse

Photodissociation: the breaking of chemical bonds in molecules whose "work function" is exceeded by high-**frequency** radiation

Photosynthesis: The process by which green plants make **carbohydrate**, CH_2O, by using high-**frequency** visible solar radiation to strip hydrogen from water and add it to carbon dioxide, releasing oxygen as a waste product

Plane of the ecliptic: the plane in which all planets orbit around Sun; Pluto, which is not a planet, has its own, distinct orbit

Polar stratospheric clouds: thin clouds that form in the lower stratosphere when the dew point is sufficiently low in mid-winter, below ~ -78 °C

Polyatomic nonmetal: elements that tend to fill their outer shells by **catenation**, including carbon, phosphorus, sulfur, and selenium

Post-transition metal: a large group of elements, including boron, B; aluminum, Al; silicon, Si; tin, Sn; and lead, Pb, which behave like metals in preferring to shed their **valence** electrons rather than **catenating**

Ppm: parts per million, used to describe concentrations of trace gases in the atmosphere

Preboreal warming: a major warming event, from 11,950 to 9,375 years ago, that ended the last ice age

Precambrian eon: that portion of geologic history before the Cambrian period, which began 541 million years ago

Prokaryotic cell: the primitive cell type of the cyanobacteria, lacking intracellular membranes and organelles, and having circular chromosomes

Reactive species: chemical species in river water whose concentrations are greatly reduced upon entering seawater, mainly by organisms

Redfield ratios: a constant ratio of carbon (106), hydrogen (263), oxygen (110), nitrogen (16), and phosphorus (1), in protoplasm from a wide variety of biological organisms found by A. C. Redfield in 1934

Resonance: that state in which a system is most responsive to vibrational forcing from an external frequency source; includes both a fundamental frequency and overtones thereof

Respiration: the oxidation of carbon compounds by an organism to obtain energy; oxygen is the most efficient electron acceptor, but nitrate or sulfate can also be used; it is an **acidifying reaction**, as it produces hydrogen ion: $CH_2O + O_2 \rightarrow HCO_3 + H^+$

Shell, electron: a region surrounding an atom within which one or more electron **orbitals** are arrayed

Solstice: one of two points along a planet's orbit where the primary (e.g., Sun) is perpendicular to the planetary axis

Sturtian glaciation: a late **Precambrian** glacial episode that spanned the 57 million year interval from 717 to 660 million years ago

Tectonic plate: one of 7 major, and several minor slabs of rigid bedrock that conform to Earth's surface, over which they can slide on the asthenosphere, a partially molten layer of the underlying mantle, as they grow by addition of erupted **basaltic** lava to their **constructional edges**

Terahertz (THz): a frequency of one quadrillion cycles per second

The Fundamental Equation of Death: a composite equation, $CaCO_3 + CH_2O + O_2 \rightarrow Ca^{2+} + 2HCO_3^-$ (the reverse of **The Fundamental Equation of Life**), combining the processes of **respiration** and **carbonate solution**; if these processes run at equal rates, they preserve the pH of the ocean; if this equation and **The Fundamental Equation of Life** run at equal rates, calcite and carbohydrate are consumed as fast as they are produced

The Fundamental Equation of Life: a composite equation, $Ca^{2+} + 2HCO_3^- \rightarrow CaCO_3 + CH_2O + O_2\uparrow$, (the reverse of **The Fundamental Equation of Death**), combining the processes of **photosynthesis** and **carbonate secretion**; if these run at equal rates, they preserve the pH of the ocean; if this equation and **The Fundamental Equation of Death** run at equal rates, calcite and carbohydrate are consumed as fast as they are produced

Theory: a statement about the operation of a natural process which is not contrary to any known hard data from the Earth system. Requires validation by hard data

Troposphere: the lowest shell of Earth's atmosphere below the **tropopause**, within which temperature falls with altitude; here convective overturn is active, causing all of Earth's weather

Tropopause: the top of Earth's **troposphere** and the base of the stratosphere at 11 mi (17 km) elevation in the tropics and 5.6 mi (9 km) elevation at the poles

UNFCCC: the United Nations Framework Convention on Climate Change, an international treaty with 197 signatories; it has guidelines, but no binding limits on greenhouse gas emissions or enforcement mechanisms

UV-B radiation: a solar radiation band from 953 to 1037 **GHz frequency** (315 to 289 nm **wavelength**); active in ozone destruction; 48 times the frequency of Earth's infrared radiation

Valence: the tendency of an atom to shed or receive electrons in order to complete a stable outer shell of 8 (or 2 in the case of hydrogen)

Wavelength: the distance between crests (or troughs) of a repeating cycle; designated by the Greek letter λ (lambda); the inverse of **frequency**, ν; $\lambda = 1/\nu$

Printed in Great Britain
by Amazon

63086512R00076